How to Think Like an Ecologist

How to Think Like an Ecologist offers a witty and engaging guide to the science behind restoring the planet and, by association, ourselves.

Aimed at non-science students, this book demystifies the basic science of environmental global changes, from climate change to biodiversity loss, while tackling the natural constraints that shape our future. The book dives into the unique power of science as a tool for uncovering truths about the natural world and equips readers with strategies to spot and counter scientific misinformation. Exploring how differing views of nature can spark conflict, it emphasises the importance of clarity and community in setting environmental priorities. To address environmental challenges requires a basic ecological literacy, so the science of ecology and evolution are examined. The text unpacks the interconnected crises we face, from plastic pollution, invasive species, ocean acidification, and more, arguing for holistic, systemic solutions. The final chapters inspire action, addressing what it takes to create a sustainable society and how to cultivate hope in these challenging times.

Filled with innovative ideas and real-world examples, this alternative textbook is a must-read for students in introductory environmental courses and those in other disciplines seeking to grasp the fundamental science behind our rapidly changing global environment.

Amy B. McEuen is a Professor of Biology at the University of Illinois Springfield, USA where she teaches courses in environmental biology, ecology, and conservation biology. She holds a PhD in Terrestrial Ecosystems from the University of Michigan, USA.

Routledge Studies in Conservation and the Environment

This series includes a wide range of inter-disciplinary approaches to conservation and the environment, integrating perspectives from both social and natural sciences. Topics include, but are not limited to, development, environmental policy and politics, ecosystem change, natural resources (including land, water, oceans and forests), security, wildlife, protected areas, tourism, human-wildlife conflict, agriculture, economics, law and climate change.

Becoming Nature Positive
Transitioning to a Safe and Just Future
Marco Lambertini, Joseph W. Bull, Leroy Little Bear, Harvey Locke, Eva Zabey, Dorothy Maseke, and Carlos Manuel Rodríguez.

The Economics of the Wildlife Trade
Theory and Implications for Conservation
Frederick Chen and Michael 't Sas-Rolfes

Rewilding and Reintroductions in Britain
Human Interactions with Conservation
Virginia Thomas

Conservation Ethics
The Challenges of Preserving Nature
Pierfrancesco Biasetti

The North American Waterfowl Management Plan
Building a Continental Conservation Partnership
Michael G. Anderson

How to Think Like an Ecologist
A Non-Scientist's Guide to Saving Our Planet
Amy B. McEuen

For more information about this series, please visit: www.routledge.com/Routledge-Studies-in-Conservation-and-the-Environment/book-series/RSICE

How to Think Like an Ecologist

A Non-scientist's Guide to Saving Our Planet

Amy B. McEuen

Routledge
Taylor & Francis Group
LONDON AND NEW YORK

Designed cover image: Getty Images

First published 2026
by Routledge
4 Park Square, Milton Park, Abingdon, Oxon OX14 4RN

and by Routledge
605 Third Avenue, New York, NY 10158

Routledge is an imprint of the Taylor & Francis Group, an informa business

British Library Cataloguing-in-Publication Data
A catalogue record for this book is available from the British Library

Library of Congress Cataloging-in-Publication Data
A catalog record has been requested for this book

ISBN: 978-1-041-21073-3 (hbk)
ISBN: 978-1-041-21072-6 (pbk)
ISBN: 978-1-003-72540-4 (ebk)

DOI: 10.4324/9781003725404

Typeset in Optima
by Taylor & Francis Books

For Leon, Auden, Nina, and all other creatures great and small

Contents

Illustrations

Figures

Tables

Boxes

Acknowledgements

It takes a village to write a book and it takes time and solitude. Thanks first to University of Illinois Springfield (UIS), and the sabbatical and awards committee specifically, for believing in this book enough to grant me a sabbatical term to write the first draft. Two of my favorite people in the world read that first draft: my partner in conservation and life, Dr. Leon Hinz, and my forever friend, playwright Thelma Virata de Castro. Particular thanks to Thelma, for asking me with a twinkle in her mischievous eye, "You don't want to write a *boring* book, do you?" That was just the voice I needed in my head as I tried to avoid using academic jargon and write this work not only for students but for present and future friends and family. I am deeply indebted to UIS art professor Brytton Bjorngaard for bringing the embedded ecosystem figure and material cycling figure to vibrant life. I'm glad the skulls make you smile, they make me smile too. Thank you also to readers who reviewed early chapters in their fields of expertise including Dr. Hua Chen for reviewing the chapter on global change. My amazing children (now adults) – Auden Hinz and Nina McEuen – continually inspire me to fight for a better world. Our kids are counting on us to get this right. My students remind me constantly that the future is in good hands. I've written it on many a whiteboard for exams and I'll write it here: you've got this! Thanks also to the Tree Advocates, particularly Jan Von Qualen, for keeping my faith in humanity alive and believing in this book even in its infancy. Jan gets an extra shout out for attempting to keep my creative use of commas under control. And finally, thanks to Earthscan/Routledge staff: editor Hannah Ferguson for seeing the book's potential and pitching it to her publisher, editorial assistant Serena Watkins for answering my questions and helping with format, and copy-editor Claire Bell for the final spruce up.

Preface

The elephant herd stands defiantly in the road as my dad slams his hand down on the horn for the umpteenth time. My stomach lurches. I am a skinny, stringy-haired kid in the backseat, ten years old, praying I live to see 11. It is 1976 and my dad is on sabbatical from San Diego State University, working as a geologist for the United Nations in Ethiopia. We live in Addis Ababa. The elephants, though, as well as me, my two sisters, my mom and dad are in Kenya. We've taken a family trip to stay at a hotel at the edge of a watering hole where warthogs come at night to drink their fill and run away with their tails in the air. My love of animals continues to run deep. Indulge me as I share just one more family memory: this time picture me red-faced as I have just passionately argued in defense of … flies. I am once again squished between my two sisters in the backseat of our small family car. My dad is driving us on a family vacation through the southwestern United States. The landscape is dotted with cacti and yellow dry ground. My family has baited me and argued that the world would be better without flies. I am filled with righteous anger in response, "What will the spiders eat?" They laugh at my angry, little-kid fury and continue to taunt, "Who doesn't want a world with fewer spiders?" Having lost the debate and my cool, I stare sullenly out the window and watch the desert roll by. Heat waves rise from the ground's surface. Fast forward to present day, more than four decades later. I have three degrees and a number of peer-reviewed publications. I teach at a small university and have received promotion through the ranks to full professor. However, I sometimes still feel like that red-faced, belligerent girl smushed uncomfortably in the backseat: equally furious with others for being so *obviously* wrong about how things work, and upset with myself because I have failed to be a strong defense lawyer for the flies.

This is a book about the state of our planet. It grew out of over two decades of teaching university courses in environmental biology, ecology, and conservation biology. My goal is to convey what I've learned along the way and to convince myself, and you, that hope is not just possible but necessary; that nature doesn't need saving, we do, and that by healing nature we heal ourselves. Scientists know a lot about how the planet works and is changing; how we've changed it. The opening chapters are about science and nature. Science is a unique way of knowing that helps us improve our understanding of the natural world. Unfortunately, we live in a time where there is widespread misinformation regarding science. We all need the tools to detect and combat misinformation because science works.

Nature is real and beautiful and messy and constantly changing. She won't end until the planet ends. She's a tough cookie. We are this big-brained, opposable-thumbed, prideful ape dangling off a recent branch of the tree of life. I am somewhat inexplicably fond of my own species. I'd really like to see us get this right. The moment we are in, however, is precarious. The branch is weighed down, bowing, and starting to creak with strain. By any metric, we are at a tipping point in our relationship with nature. We've changed the atmosphere, the oceans, the lands. We've synthesized materials that nature can't break down. These synthetic bits (aka microplastics) are in our bodies and continue to build up in living things, and are now found across the land and water. We are losing species we've never even described alongside species we have fallen in love with. The middle chapters of the book are about these changes. It's so much more than climate change. The challenge before us is stark. But nature is resilient. Anyone who thinks we can end nature has clearly never gardened. The planet will be just fine. But what about *our planet*? The planetary conditions that are conducive to all we do, can we maintain them? In the end, it's us I'm worried about. We either get it right or we don't and getting it right means accepting certain limits as real. Science, laws of physics, real with a capital R real. Sustainability is about living full, healthy, responsible lives within earth's limits. Justice means everyone having access to the incredible things our earth provides. "Hope is the thing with feathers."[1] Anyone who has caught a bird using mist-nets knows our feathered friends aren't easily caught or tagged. Nonetheless, we show up and try and celebrate when we catch one.

This book is for everyone. We are in an all-hands-on-deck situation and each one of us has a role to play in building a better future. I've tried to present the science clearly and accurately without oversimplifying. But this book is about more than science, it's also about the struggle science faces in getting heard, and how the systems we live within must be changed. I've tried to pack the book full of the voices of people who are smarter and wiser than me: Ayana Johnson, Katalin Karikó, Elinor Ostrom, Robin Kimmerer, David Orr, Carl Bergstrom, Donna Meadows to name a few. I am in awe of the community of scholars who have dedicated their lives to both understanding our current predicament and looking for a path to a more verdant future. This book is a love song to science and to those who do it well. If you, like me, are interested in finding a hopeful path to a sustainable future then follow me, your intrepid field guide, as I take you through the brambles of the scientific process, explore the swamp of science disinformation, run you through drills in basic ecology, evolution, and global environmental change, to reach our destination: sustainability and hope in the Anthropocene. Watch for snakes and snake-oil salesmen.

Note

1 The quote "hope is a thing with feathers" is, of course, the title of Emily Dickinson's famous 1861 poem (Dickinson, 1999). I love that this phrase evokes images of flight and brings to mind our feathered friends. Throughout this book, we'll be thinking about hope and where to find it. Why not start this journey by reading this classic poem? Art matters.

Reference

Dickinson, E. (1999). "Hope" is the Thing with Feathers. Franklin, R.W. (Ed.), *The Poems of Emily Dickinson*. Harvard University Press.

1 Introduction – Eight Principles that Guide Ecologists

This book is designed to have stand-alone chapters just like a standard field guide. If you see a hawk in the distance and grab a copy of *Sibley's Guide to Birds*, you're not going to start on page 1, you'll jump right to the section on birds of prey. Similarly, if you are feeling despair, go ahead and jump right to the final chapter of this book on *Hope in the Anthropocene*, or if you're curious about ecosystems, dive into the middle chapters to learn more about *Ecology and Evolution*. Maybe you are angry at a politician so the *Science Disinformation* chapter is where you'll start your journey. You may, of course, start on page 1 and diligently and methodically read each page, word by word, right through to the end. My point is just, you shouldn't feel compelled to read the book this way. That said, I have organized the chapters so that the information should flow well from start to finish.

We'll start out by asking what science is and how it is done (Chapter 2). How does science differ from other ways of understanding the world? It turns out that figuring out how the natural world works is tough but, luckily, we have a tried-and-true method that works! Planes fly. Medicines make people healthier. Science, though, doesn't happen in a vacuum. To build new knowledge, scientists need training and research needs to be funded, making science a social process. As science has developed it has revealed certain truths about our world such as the properties of certain molecules. These are facts we can't change; scientific laws we can't break. To live on the planet sustainably we need to accept and work within such natural constraints. No matter how angry it might make you, you can't stop carbon dioxide from being a heat-absorbing gas. Denial is not a pathway to sustainability. Frustratingly, we live in an age where even well-established, consensus science is often publicly doubted. This brings us to Chapter 3 which looks at the problem of disinformation and misinformation and provides tools that we all need to be able to detect and refute false claims. One thing I'll explore in this chapter is the fact that no one can be an expert in all fields. This means that science literacy involves being able to distinguish good sources of scientific information from bad. This is a book about nature and nature conservation. It makes sense, therefore, to spend some time grappling with different definitions of nature. The goal of "saving nature" is far too broad.

DOI: 10.4324/9781003725404-1

Chapter 4 delves into this issue. You'll find that, in my opinion, nature is everywhere, a fact that creates exciting possibilities for conserving wild plants and animals in unexpected places. In Chapter 5, we dig into the scientific fields that study how nature works: ecology and evolution. Ecology is the study of how organisms interact with the environment. Evolution is the change in genetics of species through time. Not surprisingly, changing environments affect the organisms that live in them, including affecting their evolution. Because of this tight relationship between these two scientific fields, I cover them in tandem. Evolution has given us a great gift: the incredible diversity of plants, animals, fungi, and microbes that run the natural world. Chapter 6 investigates this teeming world, looking at the world's **biodiversity** and how fast this diversity is declining with a particular emphasis on current rates of extinction. Specifically, we'll look at research on prior mass extinction events and ask whether humans are currently causing a 6th mass extinction. We live in an era of global change. Our species has changed aspects of the earth in profound ways. Chapter 7 examines nine environmental issues to demonstrate the breadth of the challenge before us. Some of these will be familiar, such as the climate crisis and biodiversity loss. Others may be less so: novel entities (e.g. plastics), nitrogen and phosphorous pollution, ocean acidification, depletion of the ozone layer, atmospheric change, and altering land and water. Unfortunately, the overall news is not good. Our species has pushed the planet to a new place, past what researchers have set as safe boundary conditions for six of nine planetary attributes. Chapter 7 is dense with alarming data and trends. One take away from reviewing the state of the planet is that global environmental problems must be addressed holistically. We don't want to end up playing an endless game of global whack-a-mole where we deal with one environmental challenge only to exacerbate problems for another. The search for sustainability requires big thinkers: systems thinkers. Restoring our planet to safe conditions means understanding how to affect systemic level change. Spoiler, you alone bicycling to work is not going to create the level of change needed. Chapter 8 therefore discusses systems thinking and unpacks concepts of **sustainability**. Part of the reason we are in the fix we're in is because of how global economies function so economics is briefly covered to provide insight into why economic systems have allowed such global scale problems to emerge. To solve this global crisis, there is one thing we will be desperate to find and cultivate: hope. Chapter 9 defines hope as a verb and looks at actions people around the world are taking to restore the planet. It turns out there is more than enough work to do: so, pick your passion and join the effort.

A couple of last items before you embark on your adventure! First off, you may have noticed bolded words in the previous paragraph. Definitions for all these words can be found in the glossary at the back of this book. Second, I've promised to help you think like an ecologist, to start us off there are eight core principles that I'd like you to start mulling over as you read. I will go into more detail explaining these principles when we grapple with big-N

Nature in Chapters 4 and 5. For now, I've laid out each principle with a tiny bit of justification for each. These principles also serve as a warning regarding how I view the world as an ecologist. If your initial reaction is that I'm way off – hopefully you will still tag along for the hike as I make my case in the pages that follow.

Ecologists think in systems. If you pull on one strand of an ecological web, it reverberates affecting other strands. In nature, you can never do just one thing.

Eight Ecological Principles

1. **Ecosystems are everywhere, even in places you don't expect them**

Ecosystems are systems where living organisms interact with the nonliving environment. If I ask you to picture an ecosystem you might imagine a lovely forest or a babbling brook, some place far from the hustle and bustle of daily life. The reality is, though, that you yourself contain ecosystems, no need to go anywhere at all! Your skin has living things, a microbial community, that protects it as does your gut. Take a step outside. Just one small pinch of soil from the land can contain millions of bacterial cells and miles of fungal threads. A drop of water from a pond is teaming with zooplankton, phytoplankton, and if you're lucky maybe even a water bear. These are very, very tiny "bears" so no need to worry. Everywhere there is life – there are ecosystems and much of life is microscopic.

2. **You are part of nature and nature is everywhere**

Nature isn't (just) over there, it is everywhere. You are part of nature. Nature is in our cities and our corn fields. Our economies and societies are subsystems within a pulsing ecosphere. Humans evolved on earth around 300,000 years ago, fairly recently for a 4.5-billion-year-old planet. Nature was here before we showed up and will be here after us if we go extinct. She has weathered five mass extinctions and slowly but surely built the planet's biodiversity back up after each one. Nature is patient. If we are lucky, we will learn to live within her limits and have time to do what our curious species does well: art, science, math, and a bit of frivolity.

3. **All living things exchange materials with the environment. When they die, they feed new life in a continuous cycle**

Ecosystems are always in flux. Living organisms depend on the nonliving environment and contribute back to it. You are free to think about this when you flush the toilet or just as you sit here, reading and breathing. Breathing in one set of gases and breathing out different ones. Material elements move around and around and around again and again in endless cycles. They change their location from one molecule to another via chemistry. They change location from living to the nonliving world and back. Many molecules change from solid, to liquid, to gas. Think of an element such as carbon, brought into a plant from the atmosphere as gaseous carbon dioxide (CO_2), made into a sugar through photosynthesis, then that leaf may get eaten by a bug or die and fall down to the soil. As dead stuff piles up, the carbon gets respired back into the atmosphere as bacteria breakdown the dead organic matter. Around and around the elements go. Earth's merry-go-round.

Death is necessary for ecosystems to continue running. There is a finite amount of material on earth. Pick your favorite element – carbon, oxygen, hydrogen, nitrogen, phosphorous ... There is only so much of it and we don't get any more. If there was no death, all the earth's material would accumulate in living beings and the system would grind to a halt. But there is death and decomposition of tissue provides the necessary nutrients that plant life needs to grow and start a new cycle. Death is necessary for life. (This sounds like Buddhist philosophy but it is also scientific fact.)

4. **Evolution happens and it often sucks but sometimes it's a godsend**

Adaptive evolution happens when a species' genetic composition shifts due to selection for favorable traits. It's okay if you don't fully understand what those words mean yet. I'll cover the details of how evolution works in Chapter 5. For now, let's focus on why evolution can be a major headache. This happens in cases where we are trying to control a problematic species and its numbers. There are numerous examples of species whose numbers we would like to decrease: a pest that eats crops or a bacterial strain that causes disease. Development of pesticide resistance and antibiotic resistance are two examples where, because of evolution, we enter an "arms race" with a species when we try to control it. A bacterial strain makes people sick. To try to make these people healthy, doctors give them an antibiotic that kills bacteria. Unfortunately, any bacteria that survives the medicine now has a distinctive advantage. Individual bacteria that are resistant to our medicine can survive and now reproduce, creating more and more like them, until we have a bacterial strain that does not die in response to our medicine. Fighting evolution is tough and we need to be smart regarding when, where, and how we choose to start or continue this fight. Evolution however just happens, sometimes it works against our goals *but* sometimes it works towards them. An example of a good evolutionary outcome is adaptation to climate change; many species are currently evolving to be more heat tolerant, as we make the planet hotter.

5. **Nature is indifferent**

Whole books have been written on the nature of nature. In Chapter 4, we'll see that people's concepts of nature influence their environmental behaviors. I define nature simply as the physical and biological processes that are at play in the natural world, which means: nature follows the laws and processes of physics, chemistry, and biology. We use science to understand these processes. But, nature doesn't have a care in the world; the rules and laws of physics, chemistry, and biology simply work the way they work. That nature is carefree doesn't mean that nature doesn't provide us with many gifts we should be grateful for, clearly it does. It just means nature was fine before us, is fine now, and would be fine without us.

6. **Sustainability will happen whether we are ready or not**

Sustainability is not a choice. There is no option other than living within the constraints the planet imposes.[1] You've probably seen the bumper sticker: there is no planet B. If water is used to cool the data center, it can't be used to water the crops. We live on a finite planet. We ignore the monumental challenge of creating a sustainable society at our own risk. Ideally, we can figure out how to live within the planet's means by design and in a way that fosters health and happiness. If this doesn't happen by design, it will happen by default. It is up to us to solve this riddle equitably and justly or nature will impose a solution on us.

7. **Rates matter**

Spoiler alert, I have issues with *The Lorax*.[2] This may seem off brand for a conservationist and ecologist, so let me explain. My issue with this classic tale is based on the fact that the adorable, fuzzy, orange environmentalist (i.e. the Lorax), pops out of the very first stump of the very first Truffala tree that is cut down. This suggests that a good environmentalist always opposes cutting trees, even just one. But trees are an incredible renewable resource, they provide us with beautiful wood that holds on to its carbon for a long time. Cutting trees has a place in a sustainable society. The problem with logging is not that we do it; it is *how* we do it and the rate at which we plunder forests: the *rate* at which we cut trees is what makes logging unsustainable. Forests regrow at a certain rate and they can provide us with timber forever *if* we aren't too greedy, if we take less and do things more slowly. Sustainability is often about doing activities slowly and wisely with knowledge and respect for the ecosystems involved. In many cases, we don't need to stop an activity altogether. We just need to do it differently. Sustainable use therefore means using water at a rate that the natural climate can resupply it, or, to stick with our fable, cutting Truffala trees slowly and with care for how we affect the habitat of the brown Barbaloots.

8. **Change is the one constant**

One of the challenges inherent in environmental work is working with systems that are constantly undergoing change. Change is the very nature of ecosystems and of our planet as a whole. Evolution and ecology are all about change. This means that often what we are seeking to conserve is a process rather than a particular state produced by that process. A healthy ecosystem with natural processes is the goal, rather than an ecosystem that looks like one from the past. Part of what makes our current situation so challenging is that humans have now influenced the earth to such an extent, saving things we value sometimes requires changing things even more. There are no simple answers.

Those are our eight ecological principles and I will touch on them throughout as we grapple with building a sustainable society in this new environmental era, which many are now calling by a whole new name because of the scale of changes we've made to our planet. Goodbye Holocene, hello Anthropocene.

Notes

1 If your first thought is that we don't need to live within the constraints of our planet because "we are a smart species, we'll just find another planet." I would suggest you read "Becoming Earth" by Ferris Jabr (2024). This book makes a strong case for the uniqueness of earth both in terms of its minerals and life. Jabr discusses how abiotic and biotic forces have interacted and influenced one another over billions of years. The result is a spectacular, unique planet that supports both us and other species.
2 If you want to jump ahead, I explore the well-known environmental folktale by Dr. Seuss (1971), *The Lorax*, in more depth in Chapter 8.

References

Jabr, F. (2024). *Becoming earth*. Random House.
Seuss, T. (1971). *The Lorax*. Random House.

2 The Science of Doing Science

Social Forces and Natural Limits

Why Science Matters

To live sustainably on our planet, we need to first understand how the natural world works and second accept its natural constraints. Currently, our planet is in a state that functions well for our species. Every single human enterprise is dependent on having a planet that is conducive to our needs. It is important, therefore, that we figure out how the activities of 8 billion+ people affect not only ourselves but also the millions of other species with whom we share the planet's lands and waters. We will see in later chapters that we have fundamentally changed our home planet. But how do we know this? In one word, science. Science provides us with a tried-and-true method that both increases our understanding of the natural world, and allows us to develop new tools that enhance both public and environmental health. You may think you already know how science is done: the scientific method. But science is also a social process, scientists need to be trained, research programs need to be funded. To understand this aspect of science, we'll look at the life of a phenomenal scientist who played a key role in Covid vaccine development: Dr. Katalin Karikó. In this chapter, we will follow her journey as a fascinating example of how science, including its social aspects, works and how new knowledge, and in this case new medicines, arise. This chapter also covers basic scientific laws and natural constraints, including the laws of physics and chemistry. These laws create limits to what is possible. To live well on our planet, we have to work *with* nature and she has laws no one can break. Despite the incredible power of science, it is also important to acknowledge that science is just one way of discovering truths about our world. Any time we discuss science and sustainability we must remind ourselves that we rely on other disciplines, the humanities and the arts, when we grapple with life's most difficult questions.

Prior to SARS-COV-2, and the disease caused by this virus, Covid-19, vaccines were made either of inactivated virus or from isolated surface proteins from a virus. Technically, viruses are not alive because they need a host cell to replicate themselves. They are nothing more than a little piece of either DNA or RNA wrapped up in a membrane with some surface

DOI: 10.4324/9781003725404-2

proteins embedded on that membrane that they use to gain entrance into host cells. They don't have all the cellular parts necessary to make more of themselves so they hijack (infect) other cells and use the host cell's mechanisms to replicate. You can make a virus unable to replicate, inactivate it, by heating it up which breaks it into bits. This is similar to disassembling a machine. An inactivated version of a virus is like a big pile of machine parts. The machine doesn't work but the parts can be used to show your immune system what it needs to look out for. Vaccines aren't always the whole pile of parts. Scientists can also use the genetic code for a single part, the virus' surface protein, and then use modified cells in culture to produce this protein for purification. Don't worry too much about the molecular biology of it all. What is important is the fact that vaccines prime our bodies to be on alert so if a real infection occurs our immune systems fight it off very effectively. Vaccines save lives. Having a vaccine to fight off an emerging disease sweeping the planet is an amazing tool to have in the public health toolkit. But, how, with so much at stake, does one develop a new vaccine quickly? This was the predicament the global scientific community found itself in in early 2020. It typically takes several years of development and then safety testing to bring a vaccine to market but in 2020 the world didn't have several years. Lockdowns were sweeping nations; time was of the essence. Covid vaccines were created in record time because of a global community of scientists, each working on distinct pieces of the puzzle. These vaccines were also possible because of basic scientific work that had been going on for decades before Covid hit. This work had built up our understanding of different biomolecules. The urgency of the pandemic made scientists revisit some research, only to discover a scientific breakthrough had been quietly sitting in the journal *Immunity* since the mid-2000s. A breakthrough that might help them design a vaccine in a new way. Why wasn't this breakthrough acknowledged earlier? That is a hard question to answer but part of the answer lies in the personal story of one of the paper's authors. Dr. Karikó's story is one of persistence: a refusal to give up doing her science even in the face of adversity. Science is very much a social process.

Scientists Are People Too – From Testing Hypotheses to Building Consensus

Hungarian biologist Katalin Karikó thought messenger RNA (mRNA) had therapeutic potential from early on in her career. However, the first chapters of her story catalog setback after setback: losing funding, losing jobs, papers rejected, even getting evicted from her lab space. Think of a type of academic and professional setback and it probably happened to her. Now Karikó has a Nobel Prize but that came *after* her work was pivotal to saving millions of lives. Messenger RNA is just that, a message, written in nucleotide code. The message it contains is how to make a specific protein. As you read this, tons of these little messages are scooting around in your cells, telling them what proteins need making – how to make hemoglobin or maybe more alcohol dehydrogenase (if you drank last night). Karikó knew there were hurdles to overcome if mRNA was ever to work therapeutically – two big hurdles. First, it degrades too easily – poof and it's gone. This is the nature of this molecule – it is just supposed to get its job done (tell the ribosome how to make a protein) and then go away. Also, she would find out later in her career, that when unmodified mRNA was injected into animals it caused an inflammatory response. Not something you want in any medicine.

Like any good scientist, Karikó had some ideas about how to overcome these hurdles. Her big idea – **hypothesis** if we want to get all science-y – was that mRNA could be used therapeutically, specifically she thought it could be helpful in treating conditions such as heart disease or strokes. (Reminder, definitions of all bolded words can be found in the glossary at the end of the book.) With hindsight it is interesting that, in the mid-1990s, Karikó wasn't thinking about using mRNA for vaccines although other scientists were. A vaccine that used mRNA would mean a person would essentially create their own vaccine internally by reading the mRNA's message. Instead of vaccinating a person with a virus' purified surface protein, the person's own cells would temporarily make that virus protein and this would illicit the associated immune response. To do this effectively though, you have to not trigger *too* strong a response (inflammation) and you have to get the fragile mRNA into the places in cells where messages can actually be read. Many of Karikó's colleagues felt the hypothesis had been effectively falsified (hence the poor funding). People had tried working with mRNA and it just didn't work. mRNA was too fragile and later work showed it caused inflammation to boot. Karikó was two things, though, next-level smart and dogged in her persistence and work ethic. She had been fascinated with liposomes for a while (as one is): little bubbles of water-soluble fat. Some of her initial research in Hungary was with liposomes[1] and those studies had shown these bubbles could be used to shuttle novel DNA sequences (genes) into cells, *and* that the DNA would be transcribed (read) by those cells into mRNA and even produce the gene's protein! That was DNA though, a much tougher molecule than fragile mRNA. Also, the cells this was done with were cells sitting in a flask in an incubator in a lab, not cells inside a body. But, still, liposomes as little bubble-shuttles for fragile molecules had potential. That still, however, left the therapeutic problem of inflammation.

When you hear scientists talk about breakthroughs it is easy to see why there is the lore of Newton getting hit on the head by an apple or Archimedes sitting in his bathtub and displacing water. Eureka! Scientists prime themselves for discoveries by immersing themselves in their given field: they've read all the articles, talked with fellow scientists at conferences, done experiments and reflected on the results. They are obsessed with a certain topic. Primed with thinking day and night about how a given system works, one day a thought just floats into their mind. Dr. Karikó's eureka moment sounds just like this, "In 2004, it occurred to me that transfer RNA doesn't cause an inflammatory response" (Nair, 2021, p. 3). This musing set her thinking about how types of RNA differed from one another, why would some RNAs cause inflammation and not others? How did these molecules differ from one another? This brought her to a **testable hypothesis** and associated **prediction – if** differences in the nucleotides were the reason for the inflammatory responses, **then** a modified mRNA wouldn't cause inflammation. She and Drew Weissman tested this hypothesis by creating mRNA with modified nucleotides and running the required experiments, and lo and behold, no inflammation! It was a breakthrough for therapeutics. They eagerly wrote it up. Sent it to one of the top scientific journals, *Nature*, and within 24 hours had their paper … rejected. The reviewers said the science was "incremental" meaning they thought it added to the body of knowledge but not in a substantial way. It wasn't a big enough breakthrough to warrant publication in *Nature*. Growing up in Hungary, English is not Karikó's native language; she had to look up what incremental meant. Weissman and she thought their discovery was huge, and the Nobel Prize committee in 2023 would later come to see things from their perspective but in 2004 the reviewers couldn't see it. Was this because she was a woman? Not a native English speaker? Was it because she wasn't high in the academic hierarchy, being a non-tenured professor? Was it just luck of the draw with the reviewers and the paper happened to get peer reviewers who couldn't see the work's importance? We'll never know, maybe some combination of these factors. This did not stop the science though. Karikó and Weissman published the findings in a lower ranked, smaller journal and their lives went on. Frustratingly, Karikó still struggled to do her science. Nine years after the modified RNA work was published, she lost her lab at the University of Pennsylvania but was able to get a new job in industry. In her autobiography, there is a picture of her and two other scientists at the international mRNA health conference in 2019. By the end of that year, and into the next, the whole world would be thrown on its head, and soon enough Karikó would be showered with awards. In the scramble to make vaccines for Covid as quickly as possible the groundbreaking work that Karikó had done 16 years earlier with modified mRNA would prove lifesaving.

When I read about Karikó's journey as a scientist, I am first struck by her incredible perseverance in the face of adversity. I am also fascinated with the complexity of the process of creating scientific knowledge and how our standard methods of explaining science to others often fail to capture the creativity, passion, and stubbornness that are often involved. There is a dryness, dare I say a boring or monotonous character, in how the standard **scientific method** is conventionally taught. I have experienced this in my classes. I pull up the time-honored scientific method flowchart in class and the mood shift is palpable. That clicking sound is students turning their brains to other thoughts; they've seen this all before. Memorized it for tests. The steps. Come up with a testable hypothesis, develop predictions, design a test, get results, see if the results confirm or refute your hypothesis, revise your hypothesis if necessary and run more tests. Finally, you write up your results and submit them for review by other scientists. One thing we can see in Karikó's story is hypotheses for any given scientist take place on many different levels. Their BIG idea, for Karikó this would be "we should be able to use mRNA molecules to make people healthier," coupled to littler ideas along the way – for example figuring out how to overcome mRNAs fragility or inflammation response. Science is puzzle solving and trouble shooting. There is a lot of asking why, why would this molecule cause inflammation but not another, why does mRNA breakdown so quickly? Could we design some kind of shield for it? Curiosity is a must as is tenacity. At the end, what is remarkable about the endeavor is that, if you're lucky and keep at it, you can end up learning something no one knew before. You add your tiny piece of information to the growing body of scientific knowledge. Plus, what you've learned is universally applicable! With smart phones in our pockets and medicines readily available, we have perhaps become inured to the marvel of what this ever-increasing body of scientific knowledge has made possible. The small wonders it has produced. Science's social and psychological aspects are missing from the standard flowchart. Reading other published work, collaborating with other scientists exploring a similar question, going to scientific meetings and talking with others. Getting funding and lab space and equipment so you can run the tests you want to run. These are critical parts to the process. The history of science shows us that access to such resources has not been equitable.

Although the basic reliance on gathering and assessing empirical evidence is found across the sciences, science is done differently depending on the specific field of study. This is because the objects people study differ among scientific fields and these differences constrain the type of evidence that can be collected. Is it possible to run controlled experiments? Is it ethical? Approaches to look for evidence vary. Sometimes these include designing new ways to test your ideas, including using math to see how well measurements from the world fit your conceptual model. Experiments aren't always an option. For example, think of astronomy, it's tough to move stars around or whip up a control universe. Despite this constraint, we have figured out a lot about our universe. Math helps a lot here: measure stuff, fit with equations, and test predictions. A central tenet of science is that the natural world is knowable. The puzzle pieces should fit together and make sense. The process of science is iterative because as you get new evidence, new data, you can use that data to reexamine your ideas. As the puzzle comes together, you get a better idea of what picture it is forming. Results should be reproducible and universal – an mRNA molecule should work the same way in Hungary as it works in Ethiopia. People are people. Scientific papers have detailed and extensive methods sections, excellent reading for insomniacs. Why have such excruciating detail? The goal is to explain methods thoroughly enough that another scientist can do the *exact* same thing and they should get the same results. Results from experiments or observations should not depend on who is doing the experiment. This is a key attribute of science – **reproducibility** – and it causes a stir in the scientific community when a study's results can't be replicated. If a study's results can't be reproduced it casts doubt around those initial findings. For obvious reasons, replication increases confidence. Results have in essence been double checked by an independent group. Over time as evidence accumulates, scientific hypotheses can become so well supported that they are accepted as scientific theories. **Scientific theories** have *more* evidence to support them than the initial hypotheses: heliocentric theory, gravitational theory, evolutionary theory. These theories have withstood multiple attempts to falsify them; despite all these tests, evidence keeps accumulating in their favor. Theories can actually become so widely accepted within the scientific community that they are accepted as facts; heliocentric theory would be an example of this. No scientist disputes that the earth orbits the sun. Unfortunately, this difference between what a theory means in science versus what it means vernacularly is exploited by those who seek to sow doubt about scientific findings and growing consensus. We'll discuss that more in Chapter 3.

Because testing hypotheses and theories takes time, **scientific consensus** forms slowly, emerging as evidence continues to mount supporting one hypothesis and refuting others. Take for example the hypothesis that human activities are altering the global climate. This is an incredibly difficult hypothesis to test (!) but scientists figured out how to do just that. To begin with, you can't test this hypothesis experimentally at scale. We can't make 20 planets that are all the same and randomly assign higher levels of greenhouse gases to 10 of them and see what happens over time and then compare the control and treatment planets. Experiments such as this are the best way to show **cause-effect relationships**; the researcher manipulates the factor they think is influencing an outcome, in this case greenhouse gases, and they monitor the result: climate. When you can do an experiment, you use **controls** or reference sites where conditions are left unaltered. This allows a comparison of what happens with and without the factor of interest. But such an experiment isn't possible with whole climates. You can, though, test the properties of greenhouse gases experimentally and demonstrate that yes, they do indeed absorb and reradiate heat whereas other gases do not (e.g. free oxygen). That's a simple physics experiment. The heat absorption property of carbon dioxide, methane, and other gases has been known for over a century. It's also not *too* hard to measure the increases in greenhouse gases in the earth's atmosphere. You sample concentrations from the air over time. The famous Mauna Loa station graph shows the increase in CO_2 with the annual undulation in values as the larger lands in the northern hemisphere green up each spring (absorbing CO_2) and then go dormant each fall. Getting a good *global* temperature reading is trickier because the earth is bleeping big (!) and varies a lot in temperature across its surface. You need to worry about urban heat island effects and other factors that might bias the measurement. Hard but doable. After all this work you can show the well-known graph of global temperature increasing lock step with increasing atmospheric CO_2. But even this just shows that two variables are changing in unison. Yes, greenhouse gases are increasing, and the earth is heating up simultaneously. But correlation does not require causation – how do you get around that problem? You've demonstrated mechanism. You know it is logically consistent that the gases could heat the planet given the properties the physics experiments have shown. Are there more ways to test the hypothesis? Yes, you can do this the same way astronomers test hypotheses about our universe – you develop ever more sophisticated **mathematical models** – bundles of equations and you test how good the math is at predicting new data – for example future temperatures. All of this work has taken decades and the consensus regarding anthropogenic climate change got stronger and stronger with every test. The first major study by the National Academy of Sciences was in 1979. Four decades of work, and the evidence, kept pointing in the same direction, confirming that human activities are changing the earth's climate – both its temperature and patterns of precipitation. This mountain of evidence created a scientific consensus on climate change but this consensus took decades to establish and hundreds if not thousands of scientific papers and reports. (There are a lot of sources one could cite about the scientific consensus regarding the climate crisis. One study is Carlton et al. (2015) a study which showed that anthropogenic climate change is accepted across scientific disciplines.) Because the primary culprit turned out to be fossil fuel use, disinformation campaigns were launched to spread doubt about the science, but this didn't stop the science. It did, however, affect public opinion which no doubt had a negative influence on implementing strong and timely climate policy.

Science is good at many things. It can make planes fly, cure disease, make new medicines and assess their effectiveness, it can even make chickens green and fish fluorescent by clipping in novel genes! Science, however, is silent on life's biggest questions. Why are we here? What gives life meaning? How does one live a good life? What is the right thing to do? Science can only test empirical hypotheses about the natural world. It can answer a lot of really hard questions but it can't answer *all* questions. As we tussle with thorny concepts and search for sustainability, we will need more than good science to guide us. Other ways of knowing and learning about our world include philosophy, religion, and art. These avenues can help us explore meaning more broadly. Science can demonstrate empirical truths about the world but other ways of knowing are needed to create beauty and to let us know when we're doing good.[2] Fiction writers create stories that explore the world through imagination, searching for truth in their own way. Environmental work is often value-laden and of course we need science but we also need the humanities and other ways of knowing to help us determine whether or not something is a good idea. There is the classic meme that shows the stereotypical science dude in a lab coat with a clipboard watching an egg hatch in the first panel and then a huge *T. rex* chasing the guy in the second panel. The captions read, "Science can tell you how to clone a *T. rex*. Humanities can tell you why this might be a bad idea." To solve environmental challenges, we will need all the ways of knowing. Yes, our policies need to be grounded in good, hard science. We want the technological approaches we take to work. The world works the way the world works and there is no getting around that. But for sustainability to be grounded in environmental justice, it is important for us to bring our moral compass on the journey. Also, as the phrase goes, earth without art is just "eh" so why not also make the journey as beautiful as possible.

Natural Laws and Constraints

You can break the laws of the land and even go to jail for it, but you cannot break the laws of physics. The earth has natural constraints and these need to be acknowledged and worked within as we create sustainable societies. Scientific laws do not need a judiciary to implement them; they have 100% compliance. **Scientific laws** describe consistent patterns and relationships that have been demonstrated by empirical evidence. Moti Ben-Ari describes this in his book *Just a Theory: Exploring the Nature of Science*:

> The law of gravitation is simply a formula that describes how the magnitude and direction of forces on two bodies is related to their masses and to the distances between them … Laws of nature are simply a description of what *is*.
>
> (Ben-Ari, 2011, p. 49)

The laws of thermodynamics are a particularly important set of laws to grasp for environmental science. This is because we are flat-out addicted to energy in various forms. One of the wonders of current society is abundant sources of energy to power our numerous gadgets. Smart phones, cars, planes, vacuums, computers, lighting, heating … the list goes on and on. We consume energy in lots of different forms throughout the day. Part of what is fueling the climate crisis is our seemingly insatiable hunger for more and more. The demands keep climbing as technologies change. For example, the development of artificial intelligence (AI) and its high energy-consuming data centers has caused an uptick in energy consumption. The ultimate source of energy for the planet is the sun. Even something like fossil fuels are here thanks to our sun. Photosynthesis allowed plants and phytoplankton to grow. Dinosaurs, zooplankton, and all manner of ancient creatures big and small munched on the plants and each other. Everything died and some of this dead stuff got buried under layers and layers of dirt. Heat, pressure, time. Millions of years later the earth has veins of oil, gas, and coal running through it which we can mine and combust. Energy in these fuels is in chemical form: long, long chains of hydrocarbons. When we burn these fuels, these chemical bonds are broken, and we reap the benefit of the released energy. Ideally, we do this in an **energy-efficient** manner, getting the most bang for our combustion buck. The laws of thermodynamics help us understand what the constraints are on energy efficiency.

The first and second laws of thermodynamics are 1) energy cannot be created or destroyed but can be transformed, and 2) any time energy is transformed **entropy** increases. The second law tells us that we can never have 100% efficiency when we do something like transform electric energy into kinetic energy to move our robot vacuum (or whatever gadget you like) or change chemical energy (oil) into kinetic (rolling wheels) or potential energy (batteries). The byproduct of energy transformations is heat. Heat is always leaving our planet and going into space. The more inefficient any given energy transformation is, the more entropy it creates and the more heat produced. This is the second law of thermodynamics, and you can't break it. Heat is a very dispersed form of energy, floating off our planet into space. The lower the energy efficiency of any energy usage, the more energy we "lose" as dispersed heat during a conversion. Technically, given the 1st law of thermodynamics, energy can't be created or lost but it does change form and changing from low entropy forms (e.g. lump of coal) into high entropy forms (heat) is how both laws are satisfied. Okay, so given all that, what does energy efficient really mean? Your energy-inefficient laptop will heat up more than a laptop that is energy-efficient and therefore converts more of the electricity you feed it into running the machine. One fun example of good intentions, bad outcome is the story of the snowy towns that changed their stoplights to energy-efficient bulbs only to find that this created a problem in winter when the lights no longer heated up enough to melt the snow (National Public Radio, 2009). Oops. Sometimes entropy is a good thing.

In the next chapter, we'll learn the story of King Canute which is a cautionary tale about confusing the laws people make with the laws of nature. For now, suffice it to say that our desires cannot change the laws of thermodynamics or other ways in which nature works. This stubborn fact makes me think of toddlers who get visibly upset when balls run downhill away from them or other objects fail to yield to their demands. We laugh at their tantrums because we are old enough to understand the futility of fighting physics. Methane absorbs and reradiates heat. A carbon dioxide molecule will behave the same way in the atmosphere regardless of whether it came out of the tailpipe of your car or a termite's gut. This doesn't mean we aren't smart and that we can't figure out clever ways to shift things in our favor. We improve technology all the time to increase its energy efficiency. We can and have done incredible things; all the time working within the constraints the world sets for us. Sometimes new technologies do buy us time, solving issues for the short term. For example, figuring out how to create synthetic fertilizer greatly increased global crop production, but we still live on a finite planet that works in certain ways.

What are some of the hard walls, natural constraints, we need to live with? Laws of physics and chemistry can't be broken – we can't change the nature of elements. No turning iron into gold. We can't change the properties of molecules – what chemistry they do, how they interact with one another or with heat, and so on. One hard limit comes with nonrenewable resources. Many gifts of the planet are in finite supply, and once you use them they're gone. Fossil fuels are a key example here. You only get to harvest the energy from those chemical bonds once, one combustion only. As Porky Pig says, "that's all, folks!" Another limit comes with rates of growth and reproduction. Animals and plants grow and reproduce at certain rates so it takes time for a population to recover its numbers after losing individuals. Growing seasons are different depending on where you are on earth, due to the tilt of the planet and associated seasons. There is a tendency to think that because of the increasing sophistication of our technologies that we've somehow escaped nature or that we can bend nature to our will. But in my view, our technologies just put us in a bubble, much like our economic systems. We bounce in this bubble-ship along the ocean of nature not even realizing how truly small our boat is while also failing to acknowledge that nature is what is keeping us afloat. Groundbreaking environmental books proclaim *The End of Nature, After Nature,* or *The Future of Nature.*[3] Don't get me wrong. I don't deny that our science and its technologies have allowed us to modify systems in ways that are mind-bending. Using gene editing to modify a sickle-cell anemia patient's cells and then reintroducing those cells to that person so they can now create normal blood cells. Wow, humans, impressive. Assisting the evolution of corals so they can withstand the coming heat. Nice work. Two things are true. First, that we have modified the planet in unprecedented ways and continue to develop new technologies to alter it, sometimes on a global scale. Second, that we are embedded and ultimately dependent upon a complex, ever-changing ecosphere that follows certain natural laws. We ignore these laws and their associated constraints at our own peril.

Conclusion

Science works and we will need it and much more to address global environmental challenges. Science is social. No one scientist can do science all alone. It takes a village to run a lab, test ideas, and increase our understanding of the world. We can't change the way nature works by willing it to work differently. We can though use those big brains of ours to change ourselves. This starts by acknowledging the predicament we find ourselves in – more than 8 billion people on what increasingly seems like a teeny, tiny planet. The Doomsday clock is at 11:58, we've procrastinated long enough; time to roll up our sleeves and get to work.[4]

Notes

1 When Kariko did her initial work on liposomes she was working behind the Iron Curtain in communist Hungary. She couldn't just order phospholipids from a catalog. What was to be done? Her colleague got a cow brain from a slaughterhouse and he and she spent a week extracting the fat molecules from the brain for their work (Karikó, 2024, p. 102).
2 Dr. Sarah McAnulty is a science communicator who uses provocative language to get folks interested in climate change and biodiversity loss. She runs the blog ("Do Good Shit") which shares information to foster action.
3 These are the titles of books by Bill McKibben, Jedidiah Purdy, and Elizabeth Kolbert respectively.
4 The Bulletin of the Atomic Scientists started the doomsday clock in 1947 and has reset the clock 26 times most recently in 2025 when the hand was moved from 90 seconds to 89 seconds from midnight. A visit to the website will show that climate change is one of the main risks moving the clock's hand closer to doomsday (Bulletin of the Atomic Scientists, 2025).

References

Ben-Ari, M. (2011). *Just a theory: Exploring the nature of science*. Prometheus Books.

Bulletin of the Atomic Scientists. (2025, January 28). *It is now 89 seconds to midnight*. https://thebulletin.org/doomsday-clock/.

Carlton, J. S., Perry-Hill, R., Huber, M., and Prokopy, L. S. (2015). The climate change consensus extends beyond climate scientists. *Environmental Research Letters*, 10 (9), 094025. doi:10.1088/1748-9326/10/9/094025.

"Do Good Shit" (n.d.) Retrieved September, 3, 2025, from https://www.dogoodshit.club/.

Karikó, K. (2024). *Breaking through: My life in science*. Random House.

Nair, P. (2021). QnAs with Katalin Karikó. *Proceedings of the National Academy of Sciences*, 118 (51), e2119757118. doi:10.1073/pnas.2119757118.

National Public Radio. (2009, December 16). *Energy-efficient traffic lights can't melt snow*. https://www.npr.org/2009/12/16/121498684/energy-efficient-traffic-lights-cant-melt-snow.

3 Science Literacy in an Age of Misinformation

Information literacy is an important skill for everyone to master. Science disinformation is a real and growing threat that can stall much needed environmental actions. More than ever, each of us needs the tools to recognize and refute misinformation in the fast-paced world of memes and social media. Here, I'll review work by cognitive scientists covering topics such as **cognitive biases** and the **continued influence effect**. I'll also review the techniques fact-checkers use to verify information. The five most common disinformation tactics are: false expert, false balance, repetition of falsehoods, fostering doubt, and appeal to emotion. We'll look at each of these in depth. We all rely on expert opinion so one of the key skills for both scientists and non-scientists alike is the ability to identify trustworthy science communicators. Sadly, misinformation's tentacles now have a global reach. I start this chapter by looking at this issue for my own country, the United States. But the chapter's case study covers how news organizations globally are fighting the continual onslaught of misinformation. Much like climate change and plastic pollution, misinformation is a global threat.

In 1863, Abraham Lincoln created the National Academy of Sciences (NAS) whose charge was, and in theory still is, to provide "independent, objective advice to the nation on matters related to science and technology" (National Academy of Sciences). A man before his time, Lincoln, and the congress which passed the act that established the NAS, realized that the American public needed a source of scientific information that they could trust, a source that would be independent of politics. For over 100 years, the NAS remained a strong, credible source for good science, including on politically hot issues such as climate change. Through 2023, NAS diligently published the National Climate Assessment reports that cover the climate crisis with a specific focus on expected impacts in the United States (Crimmins et al., 2023). In addition to publication, these reports and associated resources were widely available on the government's NAS website. In 2025, under the second Trump administration, however, this source of information has been suppressed (Scientific American, 2025a). The primary link to the 5th report is still on some government websites such as the USDA (as of September 2025) but the link is dead. The USDA site with the dead link to

DOI: 10.4324/9781003725404-3

the 5th climate change report now contains a warning that content is under review. This suppression of climate science in the United States was not unexpected given the appointment of Stuart Levanbach, who was a known critic of NAS reports, to oversee the very agency that produces the climate assessments (Scientific American, 2025b). The NAS climate reports were also directly targeted by Russell Vought and others in the Project 2025 playbook. Scientists' concerns over the appointment of Levanbach were dismissed. Specifically, Scott Waldman reported, "In response to questions from Politico's E&E News, a Trump administration official with the Office of Management and Budget described the scientists' concerns as "fake news"" (Scientific American, 2025b). You can still find the 5th National Academy of Sciences climate report on the web if you dig a bit but the active suppression of credible science by a government highlights the challenges science faces in the digital age. Sadly, such suppression is not unique to the United States with disinformation campaigns and associated misinformation now having a global reach. The tools taken to combat this problem are, by necessity, regionally unique (Box 3.1).

Box 3.1 Climate change disinformation: Combating a global problem

Disinformation and misinformation about climate change have a global reach. The 2023 International Panel on Climate Change (IPCC) report was the first to list disinformation as a major threat preventing climate action. Despite misinformation's global spread, the nature of disinformation: where it originates from, how it spreads, and how it is fought varies a great deal among countries and across regions. Dolores Palau-Sampio, Paz Crisóstomo Flores, and Maria Josep Picó Garcés (2024) studied more than 400 items submitted to nine fact-checking platforms across seven countries: six of these countries were in Latin America and Europe, and the seventh one was in North America, the United States. More than three-quarters of climate change information was found to be false or misleading (86%). The percentage of false content was found to be higher in the US (73%) and Europe (72%) than in Latin America (47%). In all countries, social media was the dominate avenue through which misinformation was spread (52%). The second most frequent source of misinformation was inaccurate websites (26%). Misinformation coming from political sources was greater in the US compared to the other regions. Europe had a higher percentage of misinformation originating from inaccurate websites compared to other regions. Globally, dominant narratives included attacks on environmentalists and sources of scientific information, misinformation about renewable energy, misinformation about the causes of climate change, and/or conspiracy theories.

Because of the widespread nature of misinformation across scientific and non-scientific topics, verification platforms have become a global phenomenon showing dramatic increases over the last two decades with more than

100 countries now having fact-checking sites (Schiffrin and Cunliffe-Jones, 2022). Individual countries face unique challenges when fighting misinformation. Low literacy rates and low internet access are a barrier in some countries. Multilingual societies have to deal with disseminating accurate information in many languages. To be effective in these different environments, fact-checking platforms and news organizations must use different techniques in their responses; techniques that account for unique informational, cultural, and technological situations. To spread facts in areas where fewer people have internet access, CongoCheck in the Democratic Republic of the Congo uses phone numbers so that accurate information can be distributed via text messages. In Togo, a small country in western Africa, TogoCheck sends short audiograms via WhatsApp to increase impact and accessibility to communities with low literacy.

News agencies throughout the world must similarly tailor their approaches to mesh with their individual country's populations and politics. Nabeelah Shabber, Julie Posetti, and Felix Simon (2022) conducted interviews with individuals working for three news organizations located in the Philippines, India, and South Africa. Such organizations have been targets of disinformation campaigns. One response these news organizations have taken to these attacks is to investigate the attacks themselves and report on impacts. *Rappler* in the Philippines, which was called "fake-news" in the State of the Union by President Duterte, live-streamed the arrest of its reporters, conducted investigations into the use of propaganda in social media, and developed databases to map disinformation networks. *The Daily Maverick*, in South Africa, took a page from an older playbook and started print distribution in a common supermarket chain. These news groups also hosted webinars that taught the public fact-checking and worked to strengthen their relationship to their audience to foster trust and loyalty. Just as misinformation now has a global reach, so too do fact-checking organizations. Across the planet, people are working to increase the spread of facts, good science, and evidenced-based analyses.

All of us are immersed in what is sometimes called an information "ecosystem." As you might anticipate, as an ecologist I would not call this morass of facts, fiction, posts, videos, misinformation, disinformation, and spectacle an **ecosystem**. This is, however, a word that is increasingly used in the field of communication and I guess I'm okay with it as long as it is viewed as metaphor only as we hack our way through the entangled bank. The task of finding reliable information has become increasingly demanding as information sources have proliferated and as digital sharing of information has increased. How is a nonscientist to distinguish good information from bad? Good science from pseudoscience? Tools to fight disinformation and misinformation are needed now more than ever so the beleaguered truth has a fighting chance. (Truth is a loaded word at universities; academics can debate ad nauseum how much we ever can know reality. Please be kind and allow me to use this word as a short cut for fact-based reality.[1]) Fighting disinformation is an uphill battle. Unfortunately, simply refuting falsehoods with facts and evidence is not sufficient to change people's minds. People hold different worldviews and these views influence a person's response to new information. In addition, if someone has already encountered misinformation on a topic, that misinformation can be "sticky" because of what cognitive scientists call the **continued influence effect**. People have a hard time removing early false information from their understanding of a topic. They tend to reintroduce that early misinformation in their explanations even after they have earlier agreed that the information was refuted. On top of this psychological barrier, we add politics and bad actors who are sometimes actively working to spread disinformation. Stephan Lewandowsky and others (2017) write about how we should see our current misinformation landscape as being actively used by political drivers to create a mythical or alternate reality where evidence no longer matters.[2] As you might imagine, this is a terrifying idea for scientists, given that evidence is the very basis of scientific understanding and progress. It should almost go without saying that finding ways to escape misinformation traps is critical for science literacy. Luckily there are methods to train ourselves to develop what some call an "immunity" to misinformation.

The world works in particular ways and though misinformation can be used to create a psychological alternate reality that appeals to people with certain worldviews, misinformation cannot change how natural systems work. You can't break the laws of physics, chemistry, or biology (Chapter 2). In environmental terms, species will keep evolving and the climate will still get hotter as greenhouse gases increase, regardless of whether people believe in the laws of thermodynamics, evolution, or climate change. The story of King Canute is a fun one to think about in this context. The 12th-century historian, Huntington contended that this pious 11th-century king set his throne at the edge of the sea and commanded that the tide not come in. Huntington wrote that Canute spoke the following to the incoming waves:

> "Thou too art subject to my command ... I command you, then, not to flow over my land nor presume to wet the feet or robe of your Lord." The tide, however, continuing to rise as usual dashed over his feet and legs without respect to his royal person. Then the king leapt backwards, saying: "Let all men know how empty and worthless is the power of kings."
>
> (Huntington, 1853, p. 199)

It is doubtful that this really took place but it sure makes a good story. A good enough story to have lasted centuries. We seem, however, to still not have learned the lesson which is no doubt part of why the tale persists. The lesson is that nature not only bats last, she owns the stadium.

Misinformation versus Disinformation

We've all been misinformed at times; it happens. Despite your best intentions, you tell someone something that turns out to just not be true. With the internet at our fingertips now, this usually plays out as your friend giving you a skeptical look, or writing a comment, followed by a brief exchange of differing opinions, and then a web search to get more details. When the internet was but a babe we all had such faith that it would bring on a new age of enlightenment for just this reason. All the world's information only a mouse-click away! It hasn't, however, played out *quite* that way, instead it seems as if the fight over facts gets more ludicrous with time. Do we really have to persuade people the earth isn't flat?! **Misinformation** happens. People are far from perfect and it's easy to sometimes think things that are just plain wrong. We get buried in our worldviews and we tend to stay in those views which creates among other problems, **confirmation biases**, an inclination to more readily believe information if it fits nicely into our existing worldview. Often, we share information in all innocence, doing it with the goal of helping others understand a situation or out of a desire for them to see our perspective. In the age of social media where we can share content almost instantly, we may unfortunately be sharing **disinformation** – information created with the express intent of furthering a false narrative. This is where things get complicated. A person who shares disinformation may truly be doing so just because they are misinformed. They often don't know that what they are sharing is disinformation because they don't know the details behind its creation. They don't intend to spread falsehoods, but they do. Add bots, politics, power, stir in some dark money and you can see why science literacy has become challenging in a brand-new way in the digital age. Because intentions are hard to measure and difficult to prove, I'll typically stick with using the term misinformation here, but it's good to keep in mind that industries and political entities have been shown to actively work to disinform by promoting false information that benefits them. There are many cases where researchers have done the hard work of gathering evidence of intent to misinform. One well known work would be Naomi Oreskes' (Oreskes and Conway, 2011) book on fostering doubt on the climate crisis. Unfortunately, there are many well paid people who work to disinform. Companies can hire what are called "product defense firms" who are paid to do just that, defend company products. Unfortunately, these firms hire their own scientists or set up cunningly named "research" groups to produce weak science that counters evidence that a given product or activity is unsafe (Michaels, 2020). Scientists and others have raised alarm bells on both the prevalence of misinformation and the societal risk it poses. West and Bergstrom (2021, p. 1) write, "Misinformation has reached crisis proportions. It poses a risk to international peace, interferes with democratic decision making, endangers the well-being of the planet, and threatens public health." If we are to create sound approaches to addressing environmental and other global crises, decisions need to made using the best available information. There is, however, so much information out there, much more than any one of us can possibly process, therefore we need sound methods of weeding out what Bergstrom and West (2021) not too delicately label "bullshit."

The Myth of the Independent Intellectual

We all rely on **heuristics**, mental short cuts that get us through all the decisions we have to make in a day. One small brain, no matter how hard it works, and how amazingly brilliant it may be, cannot be an expert on everything under the sun. Sorry dear reader, this includes your brain and my brain, too. Given that we use these short cuts to form and inform our beliefs, how can someone reliably navigate the science information landscape to discern fact from fiction? People who navigate information well know and do a number of things. First, they can effectively identify experts in various fields. Second, they know which information sources are reliable and trustworthy. Third, in science specifically, they know if there is a **scientific consensus** on a topic such as climate change, evolution, or vaccination. Let's start with the first of these, identifying valid experts. Expertise matters. I once introduced Dr. Don Wuebbles to one of my classes as a climate scientist – he immediately corrected me by saying he is actually an atmospheric scientist. I smiled at the time because Dr. Wuebbles is world renowned for his work on climate including being a lead author on multiple Intergovernmental Panel on Climate Change and National Climate Assessment reports.[3] He understands the science of climate change better than I ever will and yet he wanted to be clear about where the depth of his training lies. Training matters. To determine valid experts, it makes sense to examine what fields of study their expertise is in and how they've used that training in their careers. It's mind-blowing that we can model climate change as well as we can but this is the result of hundreds of people who have dedicated their lives to learning how the climate system works and how to make strong, robust predictive mathematical models. These are the people whom we should listen to when we want to know what is happening to our global climate and how it might change in the future. What complicates things is most people don't get their information straight from scientists or even straight from published scientific research or reports. Results from research are communicated to the public thorough the media (Allchin, 2022, Osborne and Allchin, 2024) and this is where misinformation, cognitive biases, and social media rear their ugly heads. In their 2024 paper, Allchin and Osborne call this public communication stage, "science-in-the-wild," an appropriate phrase as things do indeed get a bit turbulent and madcap at that point (Figure 3.1).

In Figure 3.1, two arrows point to the science consumer (you and me), one arrow is coming from the scientific community and another arrow from misinformation sources. There are also roles that are played by the media as

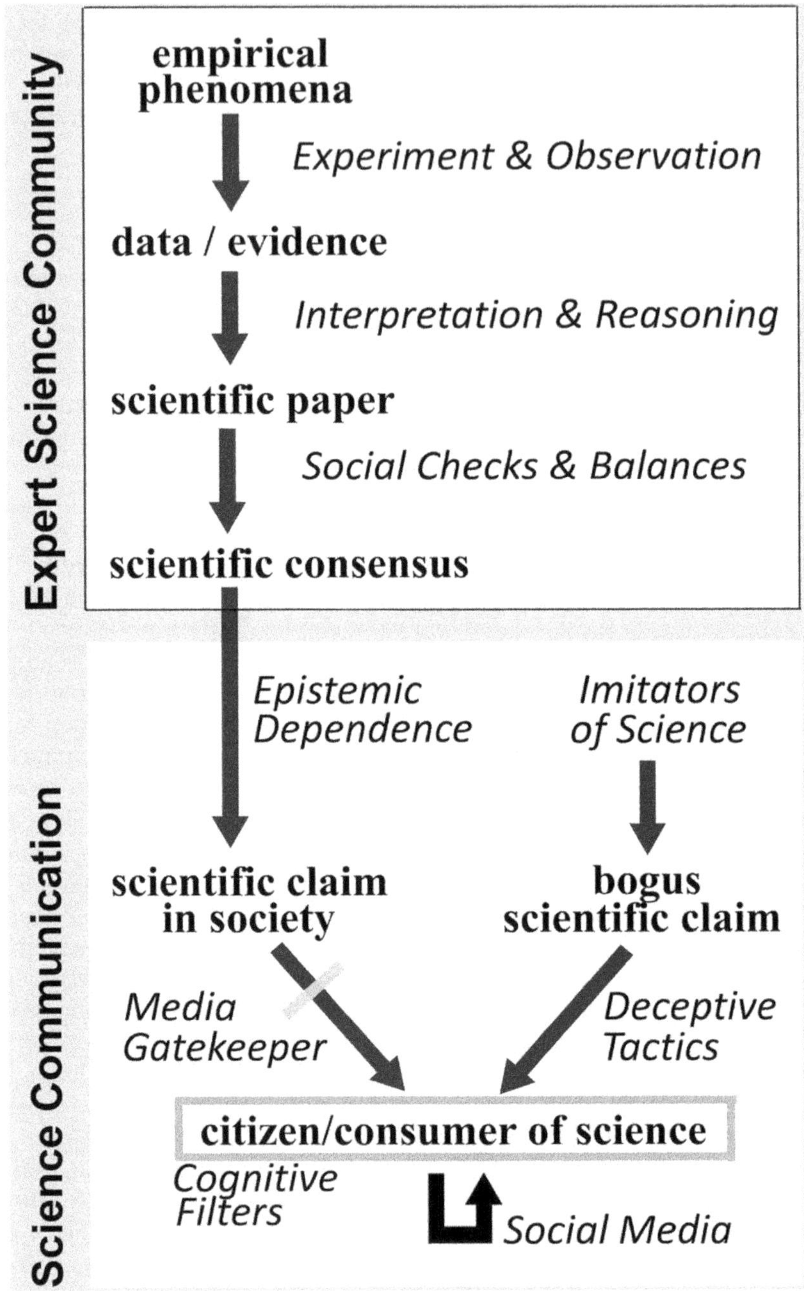

Figure 3.1 How scientific information moves

Note: Scientific information first moves through the scientific community and then into the science communication phase where misinformation (labeled here as bogus claims) also competes in informing science consumers

Source: Reprinted with permission from Allchin (2022).

a gatekeeper regarding what science is broadly communicated and the psychological cognitive filters which influence the consumer's views on any given scientific topic. In an ideal world, there would just be the one arrow with scientific information flowing seamlessly from the scientific community to the science consumer. In this ideal, questions the "science consumer" might ask would be educational and explanatory. All we would need in this scenario is good teachers, tutors, and documentary film makers, along with honest media outlets so that all science was communicated well to the consumer. The role of these groups would be to teach and communicate the science to non-scientists. We can think of this place as a world filled with earnest do-gooders with all intent to figure out how the world works, develop consensus over time, and then inform and explain the science to others. No dark money, no disinformation. Yes, evidence would change over time, as new studies were done, but scientists are always responding appropriately in this scenario; they reformulate their hypotheses in response to new results and a new consensus takes shape that explains phenomena even better and is then articulated to others. It will come as no surprise that this is *definitively not* the world we are living in.[4] The second arrow in Allchin's figure (2022, Figure 3.1) is labeled "deception tactics" and associated with this arrow comes the science consumer's associated challenge – assessment of expertise and credibility. Influencing this are the worldview of the consumer and that person's associated cognitive filters plus the influence of social media. What we all need to be good at, better at than we currently are, is evaluating expertise and information sources despite all these challenges. Thankfully, there are tools and techniques that can aid us in determining who truly speaks for the scientific community. Because of the rising challenge of misinformation, some have called for a change in science education with an increased emphasis on identifying reliable information (Osborne and Pimentel, 2022).[5]

Assessment of expertise and credibility is the first step in evaluating new information we encounter. The first question we need to ask when faced with a scientific claim is basically "who the heck are you and why should I believe you?" Or, put more politely, what are your credentials, training, and funding sources? Who do you work for? The internet is helpful in this regard because a person can quickly search for details about a given source. Let's say I pick up a book at a book sale that claims if I eat apricots every day I'll live to be 100. This book's author is Dr. Smarty Pants, Ph.D. My first question should be, Dr. Who and Ph.D. in what and from where? Is this person's training even in the nutrition field? Who does this person work for? The apricot lobby? Sometimes this important information is not even presented with the book! Red flag! Why is this a red flag? If information on the author's credentials made the author more credible, it would likely have been included in the book's author description. If we do a quick search for someone like a celebrity such as Dr. Phil we find his Ph.D. is in clinical psychology from North Texas State University and that he is currently not licensed to practice as a clinical psychologist. Information about a person is useful in evaluating the claims they make, understanding their history, training, and motivations. Stanford's Civic Online Reasoning group (Digital Inquiry Group) lists three questions to ask when confronting information online 1) who is behind the information? 2) what do other sources say? and 3) what is the evidence? (The Civic Online Reasoning group is awesome and cool and you should definitely check it out.[6]) Many people would put evaluating the evidence first in the list of tactics, but I've put evaluating evidence third simply because at some point you may hit a wall where your training and background no longer allows you to critically evaluate the evidence for a particular claim. You will have to rely on the answers you get from the first two questions – expertise, training, and corroboration from trusted sources to decide whether the information is to be trusted.

A quick example will demonstrate what I mean by hitting a wall when evaluating evidence. The following is an opening sentence from a recent peer-reviewed paper on extreme weather events. "Persistent episodes of extreme weather in the Northern Hemisphere summer have been associated with high-amplitude quasi-stationary atmospheric Rossby waves, with zonal wave numbers 6 to 8 resulting from the phenomenon of quasi-resonant amplification (QRA; Mann et al., 2018)." Hat tip to you if you can actually understand this sentence. I doubt anyone without advanced training in the physical sciences can make sense of it. The authors lost me at high-amplitude and I have no shame in admitting that I have no earthly idea what a Rossby wave is. Well shoot, and I like to think I am more qualified than many to evaluate climate science. After all, I've lectured on climate science for two decades, kept up on new findings in the field, I've even taken a course on the topic to keep up to speed. Even so, I could not have served as a peer-reviewer for this particular scientific article because I do not have the training to critically evaluate the methods the authors used or conclusions drawn from their results. But, here's the kicker, I trust the results without being able to follow the text or, dear lord, the equations that follow. Why? First, it's published in a high-quality scientific journal, *Science Advances*, published by the American Association for Advancement of Science (AAAS). I know this to be a good source of scientific content given the long history of AAAS. This association is known to publish reliable scientific content. I also know that this journal uses good scientific practices including peer-review of submitted manuscripts. This means people who do understand this science have evaluated the paper before it was published. Second, I am familiar with the lead author on this paper, Dr. Michael Mann. I know he is a professor at a reputable university, University of Pennsylvania. I know that he has multiple degrees including a Ph.D. in Geophysics from Yale which shows he has spent time getting the necessary training to develop expertise on this topic. I know that he has been researching climate extremes and climate for decades in collaboration with other well-trained scientists. You may not have known these things about Dr. Mann but, they are easy to find out. Much easier than taking all the math courses a person would need to even begin to understand all the equations in that article. Thanks scientific publication practices, thanks universities! Thanks Wikipedia and other internet sources of information. Keep up the good work.

So, we have strategies to evaluate expertise, which brings us to the second question from the Digital Inquiry Group (n.d.): what do other sources of information say about a given claim? This is where we can take a page from the fact-checkers who are excellent at what is called **lateral reading**. Lateral reading can be done both to check a given source's expertise and/or to find what other groups studying a topic say about it. The trick with lateral reading is to triangulate on any one piece of information before spending too much time with it. This is in contrast to vertical reading where you spend time reading one source in depth and try to discern if the argument that is being made is sound. Studies have shown that professors and students tend to read vertically when asked to evaluate a source whereas fact-checkers immediately work on triangulating on the source's context by searching out additional information, including corroboration from other sources. The internet is the perfect tool for lateral reading. We can look up names, news articles, find out who created a certain group, look for funding sources and so forth. If you are laterally reading, you are likely opening lots of tabs and treating each fresh piece of information like a new piece in the corroboration puzzle.

Misinformation Tactics

Although it is impossible to be an expert in every field of study, there are still many skills that each one of us can cultivate to strengthen our resistance to misinformation. These skills will also help us in discussions with others because the unfortunate truth is that the misinformed side can win a debate (i.e. persuade people of a falsehood). Cognitive scientists call developing our "misinformation muscle" inoculation – likening the tools used to a vaccine. If we train ourselves, then when the misinformation "viruses" infect our system we will be primed to resist! If everyone in a society was immune to misinformation, we could perhaps achieve the holy grail of information communication– societal **herd immunity** from misinformation. Lewandosky and Van Der Linden (2021) raised the question of whether the vaccine analogy of inoculation and immunity from misinformation could be extended to the level of a whole society. In public health, we know that when enough people have been vaccinated a population can achieve herd immunity that protects the population from future disease outbreaks. Could a society have enough people inoculated against misinformation that the equivalent would happen? In this scenario, misinformation would fail to spread because enough people recognize it and therefore do not share it with others. Stopping it in its tracks. Obviously, at present we are nowhere close to having a society with this level of misinformation awareness but we can work towards that, and the first step is honing our personal skills. One step in strengthening our response to misinformation, is learning the five major tactics that are used by those seeking to spread false information. Once you learn to recognize these strategies, you will start noticing them all over the place and with time these red flags will become easier and easier to spot. In a nutshell, these techniques include 1) fake expertise, 2) false balance, 3) fostering doubt, 4) emotional language, and 5) repetition of falsehoods (Table 3.1).

We've seen that part of being well informed regarding science requires relying on the expertise of others but to do this we need to be able to tell real experts from fake experts. Because of people's reliance on expert opinion, a common disinformation tactic is to falsely claim expertise. This means that something may look on the surface as if it is coming from the scientific community, when in fact it is not. Take for example the "World Climate Declaration" which went viral in 2022, claiming "there is no climate emergency." This document gives an illusion of expertise both by listing awards and degrees of signatories and touting the total number of signatures – more than a thousand scientists have signed! But who are the people behind this declaration and is it valid to call them climate experts? If we go to the website, the first red flag is a tab to click that says "want to sign?" Part of the reason why this declaration has so many signatures is because there is little vetting of who can sign (but they do ask you for money if you click on that tab, ha)! If you look at the credentials of those who have signed, their field of expertise and level of training is not given, just degrees and awards are listed. This is also a red flag. If people's training reinforced their climate expertise, it would be included. A little digging shows these scientists are not climate scientists and also that their funding comes from fossil fuel companies (Tigue, 2022). Unfortunately, the fact that this "petition" went viral means that damage was done by this disinformation campaign. This is true for a couple of reasons. First, the goal of these campaigns is to foster doubt about the science more than it is to completely flip anyone's opinion. Entities that push disinformation benefit from a confused, doubtful public. In this particular case, that would be citizens who are unsure of whether there is a

Table 3.1 Common disinformation tactics

Tactic	*Description*
Fake expert	Claiming or portraying sources as having expertise in an area
False balance	Insisting non-consensus opinions get equal coverage
Fostering doubt	Cultivating public uncertainty (multiple techniques)
Appeal to emotion	Fostering engagement by promising an emotional response
Repeating lies	Creating familiarity, which people associate with truth, through frequent repetition

Sources: Bergstrom and West, 2021; Cook, Lewandowsky, and Ecker, 2017; Lewandowsky et al. 2020; Lewandowsky and Van Der Linden, 2021

scientific consensus on climate change. Even if the declaration is later debunked, the doubt still lingers and affects behavior and people's priorities. This is due to the **continued influence effect** and the stickiness of initial information and early understanding related to a concept. In experiments, even when people subsequently are given *and accept* correcting information, that first piece of misinformation continues to surface later on when people are asked to make choices or explain contradictory behaviors (Osborne and Pimentel, 2022). For whatever reason, early information someone hears is very difficult to cognitively dislodge. Simply refuting that information tends to be insufficient for reversing a person's overall understanding or behaviors. Because of this damaging and lingering impact, one behavior disinformation researchers recommend we all do more of is **click restraint** which involves not sharing information until we have fact-checked it. This behavior is especially important if the information seems to confirm your worldview. We all have **confirmation biases** that make such information feel true so, perhaps counter intuitively, we need to be extra cautious when sharing material that fits with how we think the world works.

Fake expertise is just one of several tactics used in disinformation campaigns (Table 3.1). A second tactic is false balance. This tactic takes advantage of journalistic traditions that, in the pursuit of objectivity (ironically) often present "both sides" of an argument. But this "balanced" approach gives undue weight to dissenting, minority opinions and makes them seem, to the news consumer, as if a scientific consensus isn't real. This is why groups with certain anti-science agendas like to push messages such as "teach the controversy." It seems only fair! Isn't it best for our kids to learn both sides? In some cases, of course, the answer is yes. When there truly are major divisions in the science, these are wonderful opportunities for students to learn how the scientific process works. But it does not make sense to teach in this manner if one "side" is a scientific consensus and the other is propaganda or a small minority opinion. If we look again at Allchin's figure (2022), teaching the controversy is basically saying we need to teach students information from both these arrows: science and misinformation and teach them as if they are equally valid sources of scientific information (Figure 3.1). But, clearly, this isn't the case since one is coming from the scientific community and its consensus opinion and the other from those who want to muddy the waters. You can see though why false balance is effective. It gives the impression of impartiality, objectivity, and fairness. To oppose teaching something makes it seem as if scientists are suppressing views and information. The reality is, however, we don't teach that the world is flat because we have a scientific consensus that it is not.

Other disinformation tactics include fostering doubt, appealing to emotion, and repetition of falsehoods. Why is it beneficial for disinformation campaigns to foster doubt? A confused public is a desirable outcome for certain political or corporate interests. The goal is for the very nature of truth to be questioned. If there is no truth, no facts, then what basis does one have to claim that these entities are causing harm? As Lewandosky, Ecker, and Cook

(2017) explain, misinformation not only misinforms, it can make people disregard facts. Because the goal is fostering doubt, disinformation is sometimes intentionally self-contradictory. This seems wacky, but makes sense if the goal is to confuse not to refute. Dobson (2022) suggests that people move psychologically from doubt to belief and then from belief to action. So, for example, if you accept the science on climate change you might demand action from governments or industries. You might buy different products. Keeping people doubtful is therefore beneficial to the status quo. Doubt feeds inaction. Doubt can also result from another disinformation tactic: repetition of falsehoods. If a lie is repeated enough, we tend to believe it due to its familiarity. Psychologically, people tend to think familiar information is true. "Oh right, I think I heard that before." This means that making a lie familiar through repetition increases its general acceptance. This is called the **illusory truth effect**. Even if repetition of a lie doesn't lead to acceptance, it can still foster doubt. Either way the bad guys are winning. The final tactic is emotional content – promise to make someone furious, sad, elated and they are more likely to click on your link and to share the information if it indeed produces the promised emotional response. Horner et al. (2023) conducted an experiment where they showed people false news headlines and then asked what they believed and whether they would share the content. They first found, as other studies have shown, **confirmation bias**. People were more likely to believe content that aligned with their preexisting worldviews. In addition, they found that participants who reported a strong emotional response were more likely to act after reading: either by sharing the content or attempting to suppress (fact check) information. Interestingly, negative emotions were more strongly related to suppressing fake news and not spreading misinformation. Negative emotions lead to fact-checking. I guess this means there is a silver lining if new information makes you angry or sad.

Pre-Bunking, Lateral Reading, and other Life Skills

Remembering the big five misinformation tactics helps us fight the spread of falsehoods (Table 3.1), but this is not the only tool we have to slow the spread of misinformation. We can train and teach ourselves and others to be prepared for battle! Because of how our brains work, it is much more effective to pre-bunk than de-bunk. Specifically, warning someone about certain types of misinformation *in advance* works better than trying to talk them out of believing a falsehood they have already heard and believe. This is because, again, that first piece of information we get on a topic tends to be what cognitive scientists call "sticky." It's hard to dislodge misinformation once someone has already been exposed. A disease analogy works well here – it takes a lot more to fight an infection than to vaccinate a person in advance and therefore prevent the infection. How do we, then, build up our misinformation antibodies? One way is through practice and play. Several games have been developed to make misinformation vaccination enjoyable including GoViral! (Lewsey), Harmony Square (Roozenbeek and Van Der Linden, 2020), and getbadnews.com. John

Cook created the "Cranky Uncle" app which humorously teaches the user to recognize logical fallacies. If these games aren't enough, you can teach others or yourself through Stanford's Civic Online Reasoning curriculum. These materials are available free online at https://cor.inquirygroup.org/ (Digital Inquiry Group). Good habits to couple with playing these games include **lateral reading** plus **click restraint**. Check information before you share it with others. Another way to pre-bunk is to warn about misinformation on a topic. Experiments show that people who are warned about a particular tactic, say the use of false experts or sowing doubt, do better at detecting misinformation when they encounter it. Forewarned is forearmed. Another trick is to be aware and reflective about your own worldview and its associated cognitive bias. If you are politically to the left you might have a greater tendency to believe information that calls corporations into question, if on the right, you might have a greater tendency to believe information that questions the government. Knowing your own worldviews can help you develop better habits of fact-checking. Confirmation bias means we need to be *more* diligent in checking information that seems to support our worldview.

Critical Thinking and Brandolini's Principle

Critical thinking and fact-checking is all well and good but it can be time consuming and the time it takes to verify information matters. When we get to discussing sustainability (Chapter 8), you'll see I will talk about the importance of time there as well, specifically the rate at which something happens. How fast does a tree grow, what is the rate of decomposition? Recall one of the ecological principles in Chapter 1 – rates matter – how fast or slow something happens affects a system's behavior and outcomes. Bergstrom and West (2021) describe **Brandolini's principle**. This principle refers to the fact that in contrast to spreading misinformation – checking evidence, looking up funding sources, determining a proclaimed expert's training, looking into the group that creates a certain webpage – all of these tasks take time. Constructing a well-argued essay takes a lot more time than name calling. Bullying and "bullshitting" is fast and easy because you can literally just make stuff up (Bergstrom and West, 2021). In contrast, figuring out what is true requires digging, fact-checking, looking for additional sources, and, after all that legwork, *reflecting* on how what you find affects the overall strength of a claim. Douglas Allchin (2022) calls this slow method of evaluating evidence, taking an "analytical posture" – approaching new information methodically with patience, lateral reading, and click restraint. Obviously, this takes more time than just saying whatever you want to because the facts don't matter. Where did this principle get its name? Brandolini wrote "The amount of energy it takes to refute bullshit is an order of magnitude bigger than to produce it" (Bergstrom and West, 2021, p. 11). This idea has been around a while, bullshitting is fast and easy. Hobbled by our analytical posture we fear we will never win the race. Jonathan Swift wrote in 1710:

> Falsehood flies, and the truth comes limping after it, so that when men come to be undeceived, it is too late; the jest is over, and the tale hath had its effect: like a man who hath thought of a good repartee when the discourse is changed, or the company parted; or like a physician, who hath found an infallible medicine, after the patient is dead.
>
> (Bergstrom and West, 2021, p. 15)

Swift is predicting some dark outcomes from the fact that it takes time to do good things including providing evidence to support one's claims when you're correct. He would no doubt be horrified by social media. But, the "infallible medicine" that Swift discusses, although too late for the initial patient, can still help the next patient. Similarly, learning how to think critically, to spot logical fallacies, to pre-bunk and inoculate others against misinformation are all useful skills. Technology can be helpful or harmful depending on how it is used. Developing technological tools that assess information credibility and having groups that fact check, such as Science.feedback.org, saves individuals time. Once you know you can trust a given fact-checking group – you can just look up their analysis and you don't have to search for all the information yourself. This is why all of us need to take the time to find and evaluate sources of information and share those sources with others.

Conclusion

Hopefully this chapter has given you a few tools to build up your own resilience in the face of misinformation. The Debunking Handout (Lewandowsky et al., 2020) is a great read if you want more practice with the skills you've learned. As a society we have a long way to go before we are immune to the onslaughts of misinformation. As you encounter "science-in-the-wild" (Osborne and Allchin, 2024; Figure 3.1), remember first and foremost to assess the qualifications of the person or organization making the scientific claim so you can determine if they are indeed a reliable spokesperson for the scientific community. Remember the big five misinformation tactics: false expertise, false balance, fostering doubt, using emotion, and repeating lies to increase familiarity (Table 3.1). Don't forget to read laterally and to restrain from sharing information of unsure origin. If all else fails you can always befriend a scientist or two! Each of us has a role to play in decreasing the amount of disinformation and misinformation swirling around in our information ecosystems. My students tease me because my favorite meme is the "Credible Hulk" who, as one might suspect, always cites his sources (Didymus, 2018).

Notes

1 Science, as we've seen, is based on figuring out, with an open mind, what models best fit the data and being able to change those mental models as new data emerge. However, this very nuance of science has been exploited by those

seeking to misinform by suggesting there is a large amount of uncertainty when in reality the uncertainty around a concept such as gravity is incredibly small. For simplicity's sake, I will therefore use the word truth for fact-based reality with the understanding that philosophy has much more to say on the matter.

2 If you are interested in finding out more about how certain interests actively promote the spread of misinformation David Michaels examines this in his book *Triumph of Doubt* (Michaels, 2020).

3 Shameless plug for a great University of Illinois scientist: find out more about Dr. Wuebbles at his university website https://climas.illinois.edu/directory/profile/wuebbles.

4 Even Figure 3.1 oversimplifies the situation since it portrays the scientific community as only purveyors of good scientific information. The truth is much murkier because there are what David Michael's (2020) calls "mercenary scientists," scientists who are hired by groups for the purpose of producing "research" that calls into question emerging scientific consensus.

5 Currently, science education tends to focus on teaching facts, the scientific method, and how to evaluate evidence. However, this is not sufficient in the fight against misinformation because of the depth and breadth of the scientific enterprise. It is impossible for one person to be trained in every field of study and therefore be able to evaluate the evidence in every scientific paper ever written. Jonathan Osborne and others have called for changing science education to build skills in identifying reliable experts and accurate information. They argue this is a primary skill needed for life-long science literacy (Osborne and Pimentel, 2022).

6 The Civic Online Reasoning group, based out of Stanford university, provides curriculum and web exercises to strengthen your reasoning muscle. This includes videos and exercises (Digital Inquiry Group).

References

Allchin, D. (2022). Ten competencies for the science misinformation crisis. *Science Education*, 107(2), 261–274. doi:10.1002/sce.21746.

Bergstrom, C. T. and West, J. D. (2021). *Calling bullshit: The art of skepticism in a data-driven world*. Random House Trade Paperbacks.

Cook, J., Lewandowsky, S., and Ecker, U. K. (2017). Neutralizing misinformation through inoculation: Exposing misleading argumentation techniques reduces their influence. *PloS one*, 12(5), e0175799. doi:10.1371/journal.pone.0175799.

Crimmins, A. R., Avery, C. W., Easterling, D. R., Kunkel, K. E., Stewart, B. C., and Maycock, T. K. (2023). *Fifth national climate assessment*. Eds. US Global Change Research Program, Washington, D.C. USA. (doi link is currently inactivated)

Didymus, J. (2018, June, 14). *Avengers: 25 hilarious hulk memes that will turn you green*. Screenrant. https://screenrant.com/avengers-hilarious-hulk-memes/.

Digital Inquiry Group. (n.d.) *Civic online reasoning*. https://cor.inquirygroup.org/.

Dobson, G. P. (2022). Wired to doubt: Why people fear vaccines and climate change and mistrust science. *Frontiers in Medicine*, 8, 809395. doi:10.3389/fmed.2021.809395.

Henry of Huntington. (1853). *The chronicle of Henry of Huntington. Comprising the history of England, from the invasion of Julius Cæsar to the accession of Henry II. Also, The acts of Stephen, king of England and duke of Normandy*. Harvard University. Full text available at https://archive.org/details/chroniclehenryh00foregoog, p.199.

Horner, C. G., Galletta, D., Crawford, J., and Shirsat, A. (2023). Emotions: The unexplored fuel of fake news on social media. In A. Dennis, D. Galletta, and J. Webster (Eds). *Fake news on the internet* (pp. 147–174). Routledge.

Lewandowsky, S., Cook, J., Ecker, U. K. H., Albarracín, D., Amazeen, M. A., Kendeou, P., Lombardi, D., Newman, E. J., Pennycook, G., Porter, E. Rand, D. G., Rapp, D. N., Reifler, J., Roozenbeek, J., Schmid, P., Seifert, C. M., Sinatra, G. M., Swire-Thompson, B., van der Linden, S., Vraga, E. K., Wood, T. J., Zaragoza, M. S. (2020). *The Debunking Handbook 2020*. https://sks.to/db2020. doi:10.17910/b7.1182.

Lewandowsky, S., Ecker, U. K., and Cook, J. (2017). Beyond misinformation: Understanding and coping with the "post-truth" era. *Journal of Applied Research in Memory and Cognition*, 6(4), 353–369. doi:10.1016/j.jarmac.2017.07.008.

Lewandowsky, S. and Van Der Linden, S. (2021). Countering misinformation and fake news through inoculation and prebunking. *European Review of Social Psychology*, 32(2), 348–384. doi:10.1080/10463283.2021.1876983.

Lewsey, F. (n.d.) *Go Viral: Cambridge game pre-bunks corona virus conspiracy theories*. University of Cambridge. https://www.cam.ac.uk/stories/goviral.

Mann, M. E., Rahmstorf, S., Kornhuber, K., Steinman, B. A., Miller, S. K., Petri, S., and Coumou, D. (2018). Projected changes in persistent extreme summer weather events: The role of quasi-resonant amplification. *Science advances*, 4(10), eaat3272. doi:10.1126/sciadv.aat3272.

Michaels, D. (2020). *The triumph of doubt: Dark money and the science of deception*. Oxford University Press.

National Academy of Sciences. (n.d.). *About the NAS*. https://www.nasonline.org/about-the-nas/.

Oreskes, N. and Conway, E. M. (2011). *Merchants of doubt: How a handful of scientists obscured the truth on issues from tobacco smoke to global warming*. Bloomsbury Publishing USA.

Osborne, J. and Allchin, D. (2024). Science literacy in the twenty-first century: Informed trust and the competent outsider. *International Journal of Science Education*, 47(15–16), 1–22. doi:10.1080/09500693.2024.2331980.

Osborne, J. and Pimentel, D. (2022). Science, misinformation, and the role of education. *Science*, 378(6617), 246–248. doi:10.1126/science.abq8093.

Palau-Sampio, D., Crisóstomo Flores, P., and Picó Garcés, M. J. (2024). Fueling climate change disinformation: Global narratives distorting environmental risks in North America, Europe and Latin America. *Catalan Journal of Communication & Cultural Studies*, 16(2), 217–236. doi:10.1386/cjcs_00110_1.

Roozenbeek, J. and van der Linden, S. (2020). Breaking Harmony Square: A game that "inoculates" against political misinformation. Harvard Kennedy School (HKS) *Misinformation Review*, 1(8). https://nrs.harvard.edu/URN-3:HUL.INSTREPOS:37366465.

Schiffrin and, A. and Cunliffe-Jones, P. (2022). Online misinformation: Policy lessons from the Global South. In H. Wasserman and D. Madrid-Morales (Eds). *Disinformation in the Global South*, John Wiley, pp. 159–178. doi:10.1002/9781119714491.ch11.

Scientific American. (2025a, July 1). Top website for crucial climate information goes dark. https://www.scientificamerican.com/article/u-s-national-climate-assessments-website-goes-dark/.

Scientific American. (2025b, March 3). Trump official who tried to downplay major climate report now will oversee it. https://www.scientificamerican.com/article/trump-official-who-tried-to-downplay-major-climate-report-now-will-oversee/.

Shabbir, N., Posetti, J., and Simon, F. M. (2022). How three mission-driven news organizations in the Global South combat disinformation through investigation, innovation, advocacy, and education. In H. Wasserman and D. Madrid-Morales

(Eds). *Disinformation in the Global South*, John Wiley, pp. 193–209. doi:10.1002/9781119714491.ch13.

Tigue, K. (2022, August, 23). Experts debunk viral post claiming 1,100 scientists say "there is no climate emergency." Inside Climate News. https://insideclimatenews.org/news/23082022/experts-debunk-viral-post-claiming-1100-scientists-say-theres-no-climate-emergency/.

West, J. D. and Bergstrom, C. T. (2021). Misinformation in and about science. *Proceedings of the National Academy of Sciences*, 118(15), e1912444117. doi:10.1073/pnas.1912444117.

4 What Nature Are We Trying to Save?

The goal of environmental conservation is often depicted as saving nature. But such characterization begs the question: what exactly is nature? What do people mean when they say they want to "save nature" or "save the planet"? We'll find that the meaning varies depending on the person or organization involved. There are lots of different definitions and concepts of nature! Unfortunately, these different views of nature can lead to conflict. The word nature is just too darn broad a concept for appropriate goal setting. This means that people working on environmental issues need to be very precise when they articulate their goals and good at communicating with others. Goals also need to be shaped by the desires of local communities. If better conversations within communities take place early and often, policies and actions will more likely achieve their goals. The incredible thing about nature being such a broad concept is that we can totally run with it and work to restore nature everywhere! More plants, please. I am particularly in love with one aspect of nature: **biodiversity**. We'll see in this chapter that we can increase biodiversity in novel places by creating habitat: vertical meadows in cities, biodiverse roof plantings, forest buildings and more.

The Forest Preserves of Cook County were established in 1914 in northern Illinois with the goal of protecting natural areas around the Chicago region as the city continued its monumental growth. More than 70,000 acres in size, the management of these preserves has had its fair share of controversy. In the mid-1990s criticism exploded as people's views of nature collided. Prescribed burns, cutting down trees, applying herbicide to kill plants. To some, these practices were seen as critically important for restoring a severely degraded forest but to others such activities seemed the very opposite of nature protection. My sister, a business woman, lived in Chicago for a while and once asked me if the Forest Preserves were, "*Really* a forest?" Great question. What makes something real? From Pinocchio to the Velveteen Rabbit and Artificial Intelligence, this is one of life's persistent questions. Purists might argue a forest is not a "real" forest if its canopy is composed of invasive tree species, species that would never have been there centuries ago. Is a forest that looks nothing like the historical forest, still a forest? Technical forest definitions would lead us to say that a nonnative forest is

DOI: 10.4324/9781003725404-4

still a forest. Such rules look at the mature height of the trees, the overall extent of the tree canopy, and the size of the area that contains these trees. For these rules if an area's trees are greater than 5 meters tall when mature, cover more than 10% of the area's canopy, and live in an area at least 0.5 hectares then yes, Virginia, it is a forest.[1] Technical definitions, however, do not solve policy debates over what to do with local trees because these debates are rooted in different values. Chicago is by no means alone in having community members with conflicting views of nature that create differing opinions on policy and management practices. The Bay Area is similarly deadlocked in what environmental writer, Emma Marris (2016), has declared, "The Great Eucalyptus Debate." Eucalyptus trees, native to Australia, are a nonnative tree in California and, similar to the fight over tree cutting in Chicago, there is vitriolic debate around whether cutting down a California eucalyptus tree is good or bad.

Definitions of Nature

I'm trained in ecology, so, at least in theory, I work with nature. However, I am not typically asked to define this term for the work I do. It turns out this is true for the field. Frédéric Ducarme (2020) researched definitions of nature only to find that "thousands of scientists claim to work on 'nature', but none of them defines it." Ducarme also says the typical view of nature for academic ecologists is "an abstract network of many scientific objects." This makes sense; ecologists use the scientific method to answer testable hypotheses about the empirical (natural) world. However, I am not just trained as an ecologist. I also teach and work in the field of conservation biology. Since the field's formation in the 1980s, conservation biology has always been a different beast from the older academic field of ecology; from its start, conservation biology defined itself as a "value-laden science" (Soulé, 1985). Specifically, this means that conservation biologists value biodiversity and have the express goal of preserving the earth's biodiversity. It's not uncommon for debates to occur among conservationists regarding how best to approach protecting biodiversity. For example, I wrote an opinion piece defending what conservationists called "New Conservation," for the journal *Frontiers in Ecology and Evolution* (McEuen, 2014). Just like public debates, these academic debates are grounded in different concepts of nature and different value systems, often centered around people's role in nature conservation. Philosopher and ecofeminist Susan Griffin (1982) writes, "We are nature. We are nature seeing nature. We are nature with a concept of nature. Nature weeping." Similar to her, I do not see nature as separate from people. Unlike her, I am a bit of a cold-hearted Western scientist in my definition – abstract network of objects hits pretty close to home. I do not romanticize nature and I don't think something is at its "most natural" if it is in a particular state removed from human influence. It is not that I disagree with the central thesis of ecofeminism, that patriarchy and environmental destruction come from the same dark, deep root. I just see nature as *so* much bigger than anything people ever created or will create. I want to explain this view of nature in more detail here because there are other dominant views and these views heavily influence people's environmental perspectives and their willingness to embrace different environmental policies.

In Chapter 1, I laid out eight ecological principles for you to contemplate as you read; two of these evoke nature. They are 1) you are part of nature, and 2) nature is indifferent. Although I do not, many people do define nature in opposition to people. There are people and then there is nature – over there somewhere probably deep in the bushes. People who hold this world-view see themselves as outside not inside nature. Nature from this perspective is the opposite of culture, and/or the opposite of what is human-made or even human-influenced. Ideas of wilderness are particularly tied to this worldview. Interestingly, fewer people hold this view globally than you might think; this oppositional definition is a Western notion of nature, rooted in the rise of Christianity. Most world cultures as well as pre-Christian Europe define nature as all-encompassing, rooted in linguistic concepts of birth and flourishing or, in south-east Asian and Indian languages, as "that which follows rules" (Ducarme, Flipo, and Couvet, 2021). This latter linguistic concept relates to the idea of natural limits and laws that I discussed in Chapter 2. Ducarme writes that ecologists, through their scientific study of nature, have circled back to embracing this broader view of nature that was widely held in pre-Christian Europe and is still widely held globally. But the oppositional definition, nature as other, still pops up frequently in the West. For example, Bill McKibben's classic work, *The end of nature* (2006), argues that human influence is so widespread that nature without this influence has disappeared from earth. Elizabeth Kolbert (2021) in her book, *Under a white sky: The nature of the future*, doesn't go as far as McKibben to imply we've ended nature but does look at cases where we've used our smarts to change things *quite* a bit. For example, she visits the bubble barrier in the Illinois River that keeps invasive carp from reaching the Great Lakes. Historically, the Great Lakes watershed was separated from the Mississippi watershed. But in a feat of engineering, Chicagoans reversed the flow of the Chicago River to drain away from Lake Michigan and into the Illinois River. This was done in part so that Chicago's waste could flow into a watershed that wasn't the one with the lovely lakeshore property. This is just one example of largescale modification of nature. Kolbert (2021) also interviews coral biologists who are working to assist the evolution of coral to make it more heat resistant. She interviews geophysicists who are exploring technological approaches that would cool the planet by putting reflective objects into the atmosphere (**geoengineering**). I have no dispute with either of these writers in their claim that human influence is now pervasive and global, that everywhere on the planet has been changed in some way by human dominance. One only needs to look at climate change to demonstrate the truth of this; every ecosystem on earth now has a new climate because of us. But nature, as I know her and others know her, laughs when people speak of having killed her. The infamous Mark Twain refrain applies: the reports of her death are greatly exaggerated.

So, what exactly is my definition of nature? I think of nature as this global, complex physical-biological system we emerged from and, yes, now influence globally. Are there things that make our species unique? Yes, there are characteristics that make humans different from other species but having modified, our environment is certainly not one of them. All species influence their environment so the fact that we have changed things and, specifically, modified the environment to better meet our particular needs is pretty in line with the rest of nature. This idea of nature as all-encompassing arises at least in part from my training as a biologist. In fact, although he worded it much more eloquently, this view lines up well with good old Charles Darwin's definition of nature: "I mean by Nature only the aggregate action and product of many natural laws." Ducarme (2020, p. 1026) writes, "Since Darwin himself evolutionary biology has placed humankind within nature, and through the process-functional orientation of ecosystem ecology, nature is represented in terms of energy flows and materials cycling." The biological piece of this definition of nature is important. Nature is here, on earth, not anywhere else in the universe. Life is here and not elsewhere, at least as far as we know. It is so easy to take this wonder for granted but look around you and think for a minute about how gosh darn lucky we all are! Plants and animals that grow, flower, make more of themselves, that feed our bellies and delight our eyes and ears. A sun to warm our bodies. Rivers, lakes, and oceans with fish that feed us and water that refreshes us. We get to be one of the living things and we are the fortunate species that gets to *think* about what this all means. The curious species. The species that does science, makes art, builds exquisite structures, writes, sings, and dances. But any uniqueness we may have still arose from nature, and although we are in some ways special, we are still part of it and very much dependent on it. We just happen to be the opposable-thumbed, big-brained, bipedal animal so we do things a tad differently than other fellow travelers.

Anthropologists and evolutionary biologists place our species in the taxonomic genus *Homo* within the family Hominidae along with the great apes. This species, our species, *Homo sapiens,* first shows up in earth's fossil record 300,000 years ago. For a time, there were other human-like species, most recently *Homo neanderthalensis. Homo neanderthalensis* went extinct in the Ice Age between 35–50,000 years ago. The current hypothesis is that we in fact caused the extinction of the Neanderthals by being a better competitor for scarce resources (Timmermann, 2020). Some of us were definitely not too competitive with the Neanderthals, however, given the presence of Neanderthal genes in our species. But here we are 35,000 or so years later, trying to figure out the challenge of living on the planet. When we think about our own species, one thing we see, particularly in Western thought, is that we tend to be hubristic – prideful and arrogant. In particular, this characteristic shows up when we think about ourselves in relation to the other species we share the planet with. Bottom line, we tend to think we're a heck of a lot better than other animals. We definitely think we're smarter than the rest of the lot. However, the question posed by Frans de Waal's book (2016) is a good one: *Are we smart enough to know how smart animals are?* We tend to evaluate other species using tests comprised of tasks we excel at. We also like to think we were the first animal capable of doing cool things. Research, though, continues to show we aren't as unique as we are inclined to believe. For example, New Caledonia Crows are renowned for their ability to construct multi-part tools given only simple non-functional parts (Bayern et al., 2018). Not only that, some species are actually *better* than us at certain cognitive tasks. Birds that store food by burying it in multiple locations have fantastic spatial memory, better than the average person. Even farming is not unique to humans. Three lineages of insects developed agriculture millions of years before we were even in the picture. These groups are the termites, ants, and beetles (Mueller and Gerardo, 2002). Farming behaviors appeared a staggering 50 *million* years ago. Other species have even altered the atmosphere before; young earth's atmosphere did not have the high level of free oxygen (O_2) that we have today – 21% of today's atmosphere is free oxygen. Oxygen levels started to increase 2.4 billion years ago in what is known as the "Great Oxidation Event." This happened when cyanobacteria began producing oxygen as a byproduct of photosynthesis. What may be truly unique to our species is the scale and speed at which we have altered the earth. Our modification of the planet isn't just tool use, it is tool use on steroids compared to the humble crow. In this context, it's important to remember a third principle from the introductory chapter – change is the one constant. Earth has done nothing *but* change in its 4.5 billion years of existence. This makes "saving" nature hard if we define nature as earth's processes. Nature has been and always will be changing. If nature is processes and not a state, nature writ large does not require saving. Acknowledging that we belong in nature, however, is an important first step to accepting that we, like all other species on earth, must accept the limits placed on us by natural processes. Yes, okay, humility aside we're pretty phenomenal which is only more reason to get this right. Yes, we need to "save the earth" if by that we mean keeping the planet in a state that's compatible with human civilization so we can continue to do science, make art, and gosh darn it, have some fun at least once in a while.

Does Nature Need Saving?

If we accept this definition of nature – the complex biological and physical systems that make up our planet, then saving nature or saving the planet starts to sound a bit silly. Nature is, in fact, everywhere and all of our activities are embedded within it. Typically, we can't change how natural processes work, we just need to work with them. Natural processes per se don't need protection or saving but it would be nice to keep the planet in a state we can flourish in (more on that in Chapter 8 and 9). For now, let's think about what people actually mean when they talk about "saving nature." You often overhear people mention they "enjoy nature," like to spend time "in nature" or want to "protect nature." In this context, people may simply be referring to open, green spaces with less human activity. It is important that we understand what attributes each of us values in nature, because groups and individuals need to articulate what they'd like to save with precision. Precision can also help us avoid talking at cross purposes. The broad term nature actually leads us astray here because it can mean so many different things to different people. Some people may want to save a nice, quiet space to listen to birds sing under the shade of a tree. They might not care what species of bird or tree is there, they value the peaceful experience. You can see why a conflict might arise if restoration ecologists come in with their loud chainsaws or fire to restore this site to an oak opening or a prairie. Many ecologists, myself included, deeply value unique evolutionary lineages. For us, a rare plant that grows nowhere else on the planet (!) is extremely valuable, much more so than a common street tree that is found in abundance. If cutting the tree down creates habitat for this cherished, rare plant, the action is seen by people with this worldview as desirable, Lorax be damned. One person's act of protecting nature (unique evolutionary lineage – rare plant) is another person's destruction of nature (loss of green, peaceful spot with shade). Because of these conflicting ideas of nature and differences in values and worldviews, each of us needs to be better at naming the specific aspects of nature we value, and associated with that naming, be able to explain precisely what we are seeking to protect and why. It is important to create spaces in our communities where these conversations can play out and we can come to understand one another's different values and goals. Sadly, the word nature isn't going to get us where we need to go. We will need more specific terminology to express what we want to build together. A lot of nature's wonders are at tremendous risk and do need our help if they are to persist in the coming decades: wild places, diverse types of plants and animals, a stable global climate, green spaces, flowing rivers, animal migrations, forests, coral reefs, grasslands, estuaries, unpolluted land, water, and air. The more specific we can be about what we want to protect, the better we can be at setting goals to get there. We will spend less time bickering and more time saving lots of cool stuff if we know which nature each of us wants to save and why.

Great news, saving nature is not a pie! Different aspects of nature can be protected in different spaces and/or multiple aspects in the same place. There is no reason we can't have *both* restored prairies with rare and endangered species, and quiet, peaceful locations with shade. We'll find that we can create habitat in some pretty remarkable places – thus helping to keep population numbers of desirable species high. For example, we'll see that many types of insects can be protected by installing rooftop gardens. These gardens might not be easily accessible and can be, quite honestly, aesthetically questionable (i.e. *really ugly*) but the bugs love them! They are beautiful to the bugs. If we want to stop plant and animal declines and preserve biodiversity, these rooftop plantings are marvelous projects to support. We need bugs for bird food so we need rooftop conservation projects but we also need beautiful, wild places to rest and restore ourselves. Places that are easily accessible to folks in wheelchairs where the birds can sit in the trees and sing and the showy flowers can flaunt their colors. "Nature" conservation is not either/or it is both/and. Nature, as we've defined it here, is literally everywhere – we therefore have the chance to protect the things we love in unexpected places: sides of buildings, rooftops, roadsides, parking lots, yards, agricultural fields. Is there a bit of dirt there? If so, what a great place for a plant to grow. Such projects would also ideally be tailored to make sure they supply other natural goods and services that help the local community. Who doesn't want clean air, clean water, flood control, quiet spaces? The list goes on and on.

Nature's Gardeners

For millennia, certain cultures have approached natural systems with an attitude of command and control. Rivers were straightened and leveed, wetlands drained, species we didn't like were killed with poisons, prairies replaced with fields of corn to feed our livestock and ourselves. You get the picture. I am referring to Western culture in particular here and cultures which have a long and sordid history of colonization. This is not everyone. Many indigenous cultures developed sustainable practices with rules that prevented overuse and we'll look at that more later. At first the command-and-control approach only affected local ecosystems but as the human population grew, so did our technological prowess. This approach, bending nature to our will, which worked okay at first when there were fewer people and less technology, has now produced a sick and wounded planet. Polluted lands and waters, sick people, a climate spinning out of control, extreme weather events, toxic algal blooms, oil spills and more. Increasingly the consequences of this bull-in-a-china-shop approach to using earth's gifts is leaving us with a world of shattered glass. Whole cities go up in flames or are ravaged by hurricane force winds or drowned in floods. It is past time to find a better way – working with versus against natural systems. What if instead of seeing ourselves as earth's masters or rulers, we thought of ourselves as earth's gardeners (McEuen and Styles 2019)? Gardeners who have been gifted with brains that are good at figuring stuff out. Gardeners who can do science! Isn't it past time we use our minds to figure out how to live well while also living within the planet's means? When people think of gardening, they may picture a vegetable garden or perhaps a highly manicured yard with a bed of rose bushes. This is not what I'm talking about although, don't get me wrong, I do love vegetable gardens. I'm talking about viewing *all* of nature as a garden. What if we viewed *all of nature, all of the planet,* as a wild garden? A garden that sustains us and also that we work to keep healthy. Part of what we will want to garden is the wilds – areas of the planet where wild nature is protected and can just be wild, where people who value isolation, remoteness, and wild species can go for these experiences (Janzen, 1998). We can also, however, bring some of that wild into our farmlands and into our cities. Why not have peregrine falcons swooping down from the tops of towering buildings to feed on pigeons? Emma Marris (2013) started a compelling debate in the environmental field when she argued in her book *Rambunctious Garden* that we had become too constrained in where and how we approach protecting plants and animals. It's time, she argued, to think about moving plants and animals towards the poles if they are not doing well where they currently are. It's okay for ecosystems to have new types of species in new mixes that they never had in the past. Ecologists needed to think outside the box if we were to create a livable, resilient planet for us and others.

> Pikas don't care if they are in their native range or surviving climate change on a peak further north. Lots of endangered plants are happy as pigs in cities. So, yes, we can and should protect their native habitat but also – why not? – plant them in the TJ Maxx parking lots.
>
> (Marris, 2015, pp. 47–48)

Marris' argument is that if we value global biodiversity then we should do what it takes to protect it and not be caught up in our own pre-conceptions of what "saving nature" looks like. Do what works to protect and restore the things we value and do it with humility and an openness to learning. Are the Forest Preserves of Cook County a real forest? They are for the animals and plants that call the forest home. Speaking of homes, why not make our own homes, including high-rises, more nature-friendly (Box 4.1)?

Box 4.1 Case study – Moving up: Nature in the skies

An expansive view of nature allows us to capitalize on our own species' creativity. Once freed from the idea that sustainability and conservation can only mean trying to get ecosystems back to some previous "untrampled wilderness," we can start approaching the challenges we face creatively – embracing our species' skills in art and design. One example of this type of eco-creativity is the creation of vertical forests and meadows. Architects have now designed high-rise buildings that are covered with trees. The first vertical forest, Bosco Verticale, was completed in Milan, Italy in 2014. This building was designed with the idea that it would create habitat for trees, birds, and, of course, people. "In the Bosco Verticale the estimated ratio is as follows: 8 shrubs, 2 trees, 40 bushes, 30 animals to 1 person" (VanderGoot, 2017, p. 237). Creating a building that will support trees throughout its structure is not easy given that trees grow and therefore change in shape and weight over time. What trees can thrive in these situations? Botanists were consulted to select tree species and the "forest" is tended by flying gardeners who maintain the trees from outside the building using suspended scaffolding. In the past decade, several more vertical forests have been created across the globe including Tao Zhu Yin Yan in Taipei, Taiwan; The Rainbow Tree in Cebu, Philippines; The Secret Gardens in Montpellier, France; Trudo in Eindhoven, The Netherlands, and Vertical Forest Cairo in Egypt.

Of course, such innovations aren't limited to using trees and residential buildings, the built environment in general can be seen as providing potential locations for increasing plants and associated animals in the city or in rural areas. Vertical meadows use a different plant "pallet" than vertical forests, using more grasses and smaller herbaceous plants. In London and in other cities in the UK, Alistair Law's company installs such vertical plantings (Vertical meadow). These living walls enhance aesthetics and provide plant material for insects including pollinators (PBS, 2024). Particularly in cities where space is at a premium, new parks and other green spaces can be challenging to add due to real estate costs; creating vertical parks, meadows, and forests in such locations just makes sense (Nicholls-Lee, 2025). We've figured out how to stack people in cities on top of one another. Why not start stacking nature as well? Planning such innovative buildings requires interdisciplinary collaborations between engineers, ecologists, botanists, architects, and urban designers. Nature can thrive high in the sky and in unexpected places when we work together.

Isn't this approach, gardening nature, just the same old trap? Thinking that we know what is best for nature? Unlike command-and-control approaches, Marris (2015) is clear on the need for humility as we approach the god-like task of "saving the world." People who manage natural areas have learned to do what is called **adaptive management**. In many ways adaptive management follows the scientific method. For example, say you want to know how cutting trees influences a rare plant. You cut trees in some locations and not in others. You measure how the rare plant responds to the changes. You learn from your results (data) whether your idea (hypothesis), that cutting benefits the plants, is supported by the evidence. Natural systems are complex and those who manage them know that the systems still have a lot to teach us. We always need to be open to discovering new things about nature. This is truer than ever given the changing nature of the global environment.

We also need to recognize that we will fail sometimes. We don't talk enough about failure and how important it is for success. Humans are imperfect and the one thing we can count on is that we will mess things up once in a while. But, we are also clever, and particularly in high-risk situations (e.g. life or death), we've devised ways to dance with failure in order to minimize risk and optimize success. We know how to design systems that accommodate failure well. What does it mean to fail well? Here are some of the approaches that work. The first key is having redundancy in the system. This is like the old adage measure twice, cut once. Let's say we are in a group where everyone is well trained but each person still has a 1% chance of missing something. One sure way to decrease the probability of failure is for more than one person to look the final product over – with just two people the probability shifts from a 1% chance of failure to a 0.01% chance. Need an even lower probability? Have three people check it. There is also the old adage, it's not the crime it's the coverup. The psychology of failure deepens failure's negative impact. If a person feels shame, they may try to hide their failure. But this instinct prevents us from improving ourselves or the system that allowed the failure to occur in the first place. **Psychological safety** is a key attribute for any successful team or organization – creating an environment where people are as comfortable discussing failure as they are success. How can it be that the teams with the most successful outcomes report the most errors? It is not that they are more error prone, it is that they have learned that success is linked to acknowledging and reporting problems. A reported problem has a chance for solution. As we work to improve the global environment, we will have failures and setbacks. We will need to work on teams with members with very different training, perspectives, and worldviews. We don't have time to hide the failures we will inevitably make along the way. We need a robust approach that allows us to learn on the go and by doing so improve our approach with each consecutive iteration. The science of effective teamwork and psychological safety is fascinating. If you want to know more about failing well there is a rich literature to mine (Edmondson, 2023).

Conclusion

As we confront the formidable environmental challenges of our time, there is a great need to precisely define our environmental goals and understand differences in individual and cultural values. Communities need spaces where discussions around these goals and values take place. Saving nature, saving the planet, these slogans are too vague to be of much use and they may even cause undue conflict if definitions of nature lead to divergent environmental goals. Now more than ever, those who care about the environment need to be working as a cohesive team to achieve an array of goals. Goals need to be stacked if possible – can this place conserve an endangered species *and* provide the experience people want from a natural setting? Why not clean the water and sequester some carbon too? I personally deeply value distinct evolutionary lineages which is why I believe we should commit resources to protecting endangered and threatened species. Some environmentalists deeply value remote and wild places. Many are deeply committed to staving off the climate crisis or cleaning the air and water that life, including our lives, depends on. Nature per se does not need saving but biodiversity does, flowing rivers do, quiet spaces where we can hear the birds sing do. What nature do you want to save?

Note

1 Definitions of forest differ depending on the agency involved but share similar characteristics. The criteria used in the text comes from the UN's Food and Agriculture Organization (USDA Forest Service). The phrase "Yes, Virginia" comes from the 1897 editorial by Frances Pharcellus Church in the New York newspaper *The Sun* where he penned the now classic line, "Yes, Virginia, there is a Santa Claus."

References

Bayern, A. V., Danel, S., Auersperg, A. M. I., Mioduszewska, B., and Kacelnik, A. (2018). Compound tool construction by New Caledonian crows. *Scientific Reports*, 8(1), 15676. doi:10.1038/s41598-018-33458-z.

De Waal, F. (2016). *Are we smart enough to know how smart animals are?* WW Norton & Company.

Ducarme, F. and Couvet, D. (2020). What does "nature" mean? *Palgrave Communications*, 6(1), 1–8. doi:10.1057/s41599-020-0390-y.

Ducarme, F., Flipo, F., and Couvet, D. (2021). How the diversity of human concepts of nature affects conservation of biodiversity. *Conservation Biology*, 35(3), 1019–1028. doi:10.1111/cobi.13639.

Edmondson, A. C. (2023). *Right kind of wrong: The science of failing well*. Simon and Schuster.

Gobster, P. H. (2000). Introduction: Restoring nature: Human actions, interactions, and reactions. In P.H. Gobster and R.B. Hull (Eds.), *Restoring nature: Perspectives from the social sciences and humanities*. (pp. 1–19). Island Press.

Griffin, S. (1982). *Made from this earth: An anthology of writings*. Harper and Row.

Janzen, D. (1998). Gardenification of wildland nature and the human footprint. *Science*, 279(5355), 1312–1313. doi:10.1126/science.279.5355.1312.

Kolbert, E. (2021). *Under a white sky: The nature of the future*. Crown.

Marris, E. (2013). *Rambunctious garden: Saving nature in a post-wild world*. Bloomsbury Publishing USA.

Marris, E. (2015). Humility in the Anthropocene. In B. A. Minteer and S. J. Pyne (Eds.). *After preservation: Saving American nature in the age of humans*. University of Chicago Press.

Marris, E. (2016, November 30). The great eucalyptus debate. *The Atlantic*. https://www.theatlantic.com/science/archive/2016/11/the-great-eucalyptus-debate/509069/.

McEuen, A. B. (2014). Embrace new conservation. *Frontiers in Ecology and the Environment*, 12(6), 322–323. doi:10.1890/14.WB.009.

McEuen, A. B. and Styles, M. A. (2019). Is gardening a useful metaphor for conservation and restoration? History and controversy. *Restoration Ecology*, 27(6), 1194–1198. doi:10.1111/rec.13026.

McKibben, B. (2006). *The end of nature*. Random House Trade Paperbacks.

Mueller, U. G. and Gerardo, N. (2002). Fungus-farming insects: multiple origins and diverse evolutionary histories. *Proceedings of the National Academy of Sciences*, 99(24), 15247–15249. doi:10.1073/pnas.242594799.

Nicholls, L. (2025). 'A home for trees and birds and also humans': How high-rise forests can transform city life – and also make us happier. British Broadcasting Company. https://www.bbc.com/culture/article/20250602-how-high-rise-forests-can-transform-city-life-and-make-us-happier.

Public Broadcasting Company. (2024, March 14). *Vertical meadows: Wild hope*. Nature video series https://www.pbs.org/wnet/nature/vertical-meadows-wild-hope/29996/.

Soulé, M. E. (1985). What is conservation biology? *BioScience*, 35(11), 727–734. doi:10.2307/1310054.

Timmermann, A. (2020). Quantifying the potential causes of Neanderthal extinction: Abrupt climate change versus competition and interbreeding. *Quaternary Science Reviews*, 238, 106331. doi:10.1016/j.quascirev.2020.106331.

USDA Forest Service. (n.d.). *Forest Definitions Applied in Reporting*. https://research.fs.usda.gov/programs/fia/forestdefinitions.

VanderGoot, J. (2017). *Architecture and the forest aesthetic: A new look at design and resilient urbanism*. Routledge.

Vertical Meadow (n.d.) Bringing biodiversity back into our cities. https://www.verticalmeadow.com/.

5 How Nature Works

Ecology and Evolution

Ecology and evolution are the scientific fields that study the changing natural world. Ecologists research plants and animals and their interactions, both with one another and with the nonliving environment. As environments shift, the genetic makeup of species also shifts: **evolution**. Buckle your seatbelts, this chapter is going to be a whirlwind tour of *both* these scientific fields – approaching **ecology** and evolution in tandem due to their interconnected nature. Without getting too bogged down, I'll lead you through basic principles and terminology with a focus on the complexity and embedded nature of earth's ecosystems. Ecosystems are everywhere and they are complex systems! Ecologists would call the world's cities urban ecosystems. In this chapter, we'll look at some examples of evolution, including how an animal that lives within one Australian city – the eastern water dragon – is evolving in response to city life.

It's a warm, sunny April day, two days after Earth Day, and one day before Arbor Day. I'm sitting in a Head Start classroom in my hometown of Springfield, Illinois watching three older women from the Illinois Audubon Society act out Shel Silverstein's *The Giving Tree* (Silverstein, 1964). The biggest hit is the woman in the hand-made tree costume complete with felt leaves and apples and wrap around bark fabric. "That's not a tree, that's a person!" proclaims a girl whose braided hair is adorned with brightly colored beads that swing across her shoulders. Forty or more pre-school kids wiggle in lines on the floor, trying to keep their young bodies in crisscross, apple sauce. *The Giving Tree* is a sad tale in many ways. The tree just keeps giving and giving until all it has left to give is a stump for the old man to sit on. "What do we say when someone gives us a gift?" prompts the teacher. "Thank you!" The kids sing out. There are 12 classrooms so in total four performances. By the fourth, I'm thinking that the boy in the story never says thank you himself, never gives the tree anything. He just takes, and takes, and takes. That boy needs to treat that tree better.

In the film version of Margaret Atwood's *The Handmaid's Tale*, something bad has happened to the earth (Atwood, 1985). The specifics are sparse. Human fertility has obviously been negatively affected. Characters in the film version mention climate change and ecological collapse. I'm hearing that

DOI: 10.4324/9781003725404-5

phrase more and more in literature and film. Ecosystem collapse, ecological collapse. I'm not sure what it means other than something very bad has happened to the earth. I want to translate these fears into the science I know. Are these people talking about a loss of **ecosystem services**? **Ecological regime shifts**? Or is the fear just more general? A foreboding. The feeling that you can only take and take and take for so long. If you don't ever give back, eventually *The Giving Tree* will have nothing left to give you.

Ecology is the study of flux. Ecosystems do nothing *but* change and they can change dramatically sometimes over very short time periods (**regime shift**). Regime shifts happen when factors keep pushing on a given ecosystem, building up pressure on it until, much like the roller coaster cart that has reached the top of the track, it's time for a dramatic change. What was once a clean deep lake with a healthy lake bottom is now a mucky, green, stinky lake where no one wants to swim. Ecosystems have always changed and have always gone through these types of shifts in response to changing environmental conditions. They changed in the past. They will in the future. This is their very nature. They have as hard a time sitting still as a preschooler right before recess. Ecosystems can also shift dramatically from a state that we find desirable (a clear lake that is great for recreation) to undesirable (the same lake with a toxic algal bloom where it is unsafe to swim or even use as a water source). In the case of the lake, we have altered the system in ways that cause the shift, specifically we fertilize the land and these fertilizers make their way to our creeks and lakes. To understand these types of shifts,[1] why we see them, and how we can work to keep ecosystems in states we prefer, we need to take a step back and take a look at the science of how nature works. Two fields within biology dominate this science: ecology and evolution.

One of the ecological principles, laid out in Chapter 1, is just that: ecosystems change. Do we change them? Yes. Did they change on their own long before we were even around? Also, yes. Of the eight principles, five will guide us in thinking about how nature works:

1 Change is the one constant
2 Ecosystems are everywhere, even in places you don't expect them
3 You are part of nature, and nature is everywhere
4 All living things exchange material with the environment. When they die, they feed new life in a continuous cycle
5 Evolution happens and it often sucks but sometimes it's a godsend

Defining Ecology and Evolution

Ecology and evolution are deeply intertwined fields of study. As the environment shifts, some forms of life do better than others changing the surrounding environment as they grow in number. These forms may or may not do well in the new environment they helped create. Perhaps now they decline in number or move. Change is the one constant for both **ecosystems** (the basic unit of ecology) and **genes** (the basic unit of evolution). To quote the ancient Greek philosopher Heraclitus, "No man ever steps in the same river twice, for it is not the same river and he is not the same man." As an interesting aside, this quote actually comes from Plato who says that Heraclitus said this. We know Heraclitus believed in flux and change but we don't know how close he came to talking about a river changing (Stanford Encyclopedia of Philosophy, 2023). Regardless, any good stream ecologist would confirm the truth of the first part of this quote (rivers constantly change), and any psychologist who studies how spending time in nature impacts mood would confirm the second part. Nature changes you. Let's picture a river in flux. The amount of water flowing through a river shifts with spring flooding and summer droughts, rushing water moves rocks and sediments, water temperature fluctuates, the chemical composition of the water shifts, oxygen levels vary as does the amount of dissolve organic carbon, change is indeed constant.[2] All of these environmental changes influence which invertebrates can live there, what fish we find there, and so forth. The species in the river (fish, worms, crayfish, dragonfly babies) affect the nonliving elements (fish poop happens and it matters). Anthropogenic climate change as well as other global changes (Chapter 7) have only accelerated the rate of change experienced in nature. Perhaps this is where concern about ecological collapse comes from, the fact that we are changing the global environment at such scale with such speed. Racing towards an altered unknown world.

A typical textbook will define **ecology** as the science that studies the interaction of living organisms with the nonliving environment. You may have memorized a definition similar to this for some test in your past. But does this definition really help us understand natural systems, the field of ecology, or what future ecosystems might look like? This definition doesn't, for example, capture the very dependence of life on the nonliving (**abiotic**) environment. Interaction is doing a whole lot of heavy lifting in this definition. Let's look at some specifics, interacting with the environment turns out to include bleeding, breathing, pooping, sweating, urinating, eating, drinking, killing, dying, and last but by no means least, decomposing. This list is not exhaustive by any means. Ecologists study predators chasing prey, plants producing toxins to ward off herbivores, flowers that look like bees so that real bees will check them out and move their pollen. Fish that change sex (just for fun you should look into the biology of clownfish). What word can capture all this? Interactions yes, but also, these fantastic and fantastical organisms that are doing the interacting (Figure 5.1). This menagerie that Charles Darwin called "endless forms most beautiful." One suspects, more poetry and less scientific jargon may be needed to portray the field of ecology accurately.

Ecology is happening all the time, everywhere we look. I've said it once and I'll say it again: ecosystems are quite literally everywhere. Ecology

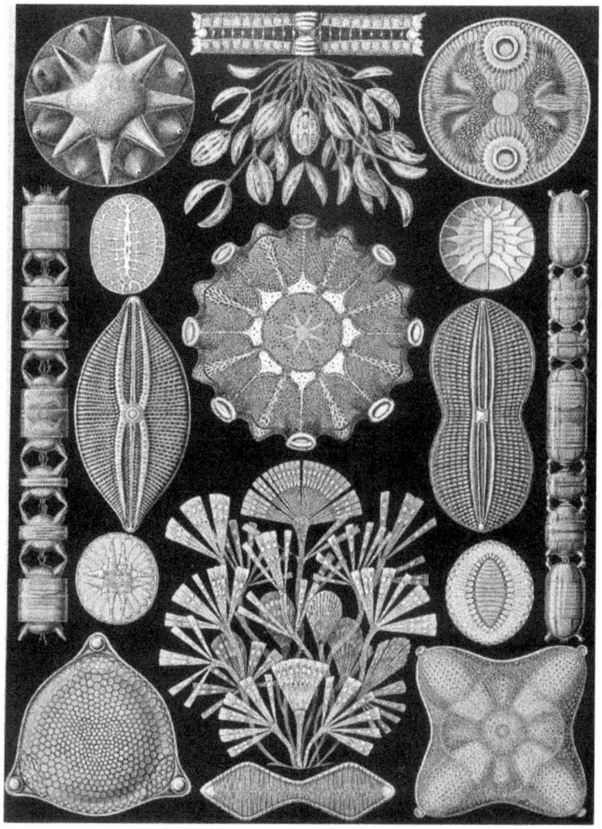

Figure 5.1 Sketch depicting the beauty and variety of diatoms
Note: Diatoms are single-celled microalgae that are found in oceans, rivers, and lakes.
Source: Haeckel (1904).

doesn't just happen down the street at the local park. It's happening right now, inside you; your intestinal track, is a teeming ecosystem chock full of a wide variety of **microbes** that keep you healthy. Microbes are a group of organisms that tend to get a bad rap. We often only hear about bacteria or viruses when they cause disease. We take for granted the hundreds of others that are either benign or extremely helpful and necessary. These different species of bacteria, viruses, eukaryotes in your gut are an example of what ecologists would call a biological **community** (lots of different **species** interacting with one another in the same place and time). Your gut ecosystem is thought to contain up to 500 species and to even influence your mood and mind (Castells-Nobau et al., 2024). What?! Science is wild. Also, this diverse microbial world is still relatively unexplored. Scientists are just now realizing how truly important these tiny species (microbes) are to just about everything. Say you want to restore a grassland on an old agricultural site, don't forget the soil bacteria and fungi if you want your restoration to be successful. The grassland will host a greater diversity of plant species and the animals if you also add the native soil microbes. Many ecosystems are invisible to the naked eye (Figure 5.2). Yes, the forest park down the street has a biological community with its grass, shrubs, trees, soil, earthworms, air, mice, stray cats, squirrels, insects, deer, and maybe foxes or coyotes (all those species living and interacting in the same place) and an ecosystem (all those species interacting with the abiotic environment) but look within the soil and within the pond, on the skin, in the guts. These tiny ecosystems within ecosystems contain hundreds more species (Figure 5.2). So, ecology is the discipline that studies ecosystems, and ecosystems are all over the place, even in your eyebrows or under your fingernails when you get done gardening. Figure 5.2 shows how ecosystems are nested within one another. One of the ecosystems drawn is the mite-person ecosystem. Most of us have these tiny

Figure 5.2 Three different ecosystems
Note: Ecosystems come in different sizes and, because of the diversity of the microbial world, are embedded in one another. Three different ecosystems are shown here: 1) plant, person, soil, 2) gut microbiome, and 3) mite-person.
Source: Original art by Professor Brytton Bjorngaard.[3] Open Educational Resource.

commensal species of mites that inhabit our eyebrows and feast on dead skin. One fun fact about these creatures is that they don't have anuses so the just fill up with poop throughout their lives. You don't want to know what happens when they die, trust me.

This complex array of species is constantly interacting with one another and with the air, soil, and water. They are also changing with each generation. The simplest definition of **evolution** is change in genetic composition over time. A quick refresher course in genetics. Living organisms pass their deoxyribonucleic acid (DNA) to the next generation. They can do this in a tremendously diverse manner of ways. It might be as simple as a single-celled organism dividing itself into two cells. It might be something more complicated like a dandelion producing seeds that are clones of the parent plant (mom) that float away sometimes carrying a child's wish with them (asexual reproduction). It could be sexual reproduction, requiring a mate, but in some cases that "mate" might be on the same individual such as different sexes of flowers on the same plant. Bottomline, there are lots of ways to pass those genes along to the next generation. The important thing is genetic information gets passed forward to offspring over time. Biological processes alone can change the genes between a parent and their offspring. Copying the DNA isn't a perfect process, so small changes happen. In species that have two or more copies of genes, a process called **recombination** can result in new genetic combinations in offspring. So just the steps needed to produce the next set of kids will bring about evolution. You might say, "I thought evolution had to do with adaptations and black moths surviving better on polluted trees than white moths and passing their genes on to the next generation so the next generation has more black moths, and so forth." You are right it does mean that *as well*, this is **adaptive evolution** the result of **natural selection**, and it is the type of evolution we talk about the most in biology, but evolution is also much, much more than that. It is ANY change in genetics across generations, not just changes in the genes that code for adaptations. The change doesn't have to be adaptive it can be any old change in the nucleotides and associated genes. In fact, most evolutionary change is thought to occur simply by what is called genetic drift. This is just the change that is produced over time by stochastic (chance) events.[4]

Ecology and evolution intertwine like jungle vines. The abiotic environment shifts as species interact with it and this changing environment creates conditions that favor some individuals within a species over others. If we focus just on adaptive evolution for now, we can see how processes work together to change species over time. The environment shifts, perhaps droughts become more frequent due to a new and drier climate, in this new climate individual plants that have long, deep roots do better than others with shallow roots. The long-rooted plants are able to access the scarce water supplies deep in the soil. All of this is ecology; plant roots interacting with the soil and water, individual plants surviving (or not), growing (or not), and reproducing (or not). The outcome though, which you may have anticipated, is evolutionary as well. Across generations we see genetic change in this plant population. The plants with the genes that encode for long, deep roots survive, the plants with shallow roots die. The surviving plants produce lots of seed the next year when the rain returns and now the plant population has changed. When the seeds of the new generation sprout, more of the resulting adult plants have long, deep roots just like their parents. If we sample this plant population at the genetic level we find a shift towards more individuals with genes that lead to deeper roots. So, the ecological conditions (the environment) influenced the plant's evolution. The plant's evolution may also have been affected in non-adaptive ways. Perhaps a shopping mall gets put in and half of the plant population is dug up and paved over, resulting in the loss of some unique genes. This is also evolution and as you might guess, we influence evolution of other species continually. Species are evolving, both adaptively and otherwise, in response to the dramatic changes we've created across the globe (Box 5.1).

Box 5.1 Evolution in cities: There be dragons

Evolution is happening all around us. We don't have to dig up fossils to find evidence of evolution. As we modify the environment, the genetics of plant and animal populations shift. Cities create a dramatically different landscape from natural ecosystems. They are filled with light and noise. They are dominated by roads, parking lots, and sidewalks (collectively called **impervious surfaces** due to their inability to absorb water). Roads cut through natural systems and separate populations of plants and animals. City parks are isolated from one another like islands in a sea of pavement. It is not surprising that plants and animals change in this environment. Remember not all evolution is adaptive, if populations are isolated from one another their genetics can change because of limited to no movement among groups or chance differences in the genetics of the initial populations. Because of all the factors that can affect evolution, scientists who study it have developed a variety of methods to measure and compare genes among populations to give them information on processes. Littleford-Colquhoun and others (2017) did this for a native lizard in Australia: the eastern water dragon. They compared ten different lizard populations – four in isolated city parks within Brisbane, and six in both isolated and connected natural areas. They caught lizards using wire nooses, took samples from their tails to get DNA, and measured features of individual lizards including hind leg length. They found that the city park populations showed greater differences in genetics among them as well as greater differences in morphologies than populations outside the city. Their study hinted at possible adaptations to park characteristics: it turns out that different types of lizard legs work better on different types of surfaces. Whether or not lizards showed adaptations, they definitely showed rapid evolution with the authors concluding, "our results suggest that city park 'archipelagos' within a single urban center could represent theatres for evolution."

Evolution in cities is by no means limited to tiny Australian dragons. Johnson and Munshi-South (2017) reviewed the now extensive evidence of urban evolution, including adaptation to city life, among wild plants and animals. German cockroaches develop glucose-aversion in response to sugar baited pesticides. Mutation, the slowest process affecting evolution, is higher in gulls in polluted industrial cities compared to non-industrial cities or rural areas. Salamanders in cities have lower genetic diversity than their country cousins. House finches develop longer and stronger beaks to crunch sunflower seeds. A plant in France has seeds that don't disperse and therefore stay within the small, isolated city population. If you listen hard you can hear the buzz and chirp of evolution all around. Change is the one constant.

How Ecosystems Work

Regardless of whether we are discussing the microbes in your gut or the birds in a rainforest, all creatures and the ecosystems they are part of share certain characteristics. One characteristic of ecosystems is flow and cycling of energy and matter, respectively. Every ecosystem is ultimately dependent on the sun to provide energy to the **primary producers**. Most primary producers on our planet are organisms, such as plants, that can take simple nutrients, light, and carbon dioxide and produce sugars. **Photosynthesis** is a powerful process and it fuels the food chain. Once those sugars are made by the producers they can be consumed by other organisms in the system. Fascinatingly, not all primary producers use light to grow, some species use chemical energy, a process known as chemosynthesis. The best example of this is the bacteria that are the base of the food chain in deep-sea hydrothermal vent ecosystems (Natural World Facts, 2022). Whether it's a deep-sea vent or your backyard, those primary producers aren't safe – they get eaten by other species. Species that derive their nutrition by consuming other species are called just that: **consumers**. As you move up the food chain, these consumers include primary consumers (eat the producers), secondary consumers (eat the primary consumers), tertiary consumers (eat the secondary consumers), and so forth. Food chains tend to be longer in aquatic ecosystems than land-based ecosystems. This is because in water there is a dominance of teeny, tiny producers, primary consumers, secondary consumers: phytoplankton or zooplankton (microscopic algae, crustaceans, protozoans and so forth). If a person eats a tuna fish sandwich they are therefore eating much higher in the food chain compared to when they eat a hamburger. A veggie burger would put them even lower in the food chain. Ecologists refer to the food chain's steps as **trophic levels**. Let's look at these two meals a person might have and think about them in terms of trophic levels. For a hamburger, corn or another grass had to be grown to feed the cow (corn is the primary producer – first trophic level). The cow then ate the grass (primary consumer – second trophic level). Finally, you grill up your burger and eat the cow (you're a secondary consumer – third trophic level). If you choose to eat a veggie burger instead, you would be eating lower on the food chain (you would then be a primary consumer – second trophic level). Now let's look at a tuna sandwich, instead of grass the base of the marine food chain is single-celled algae (phytoplankton),[5] these are eaten by zooplankton which are eaten by little fish that eat plankton which are eaten by bigger fish until we finally get up to the tuna. If you count the steps, you'll see the tuna can easily be on the 5th trophic level, which would put you on the 6th trophic level when you eat it! These long aquatic food chains have consequences for public health because fat-soluble toxins such as dichlorodiphenyltrichloroethane (DDT) or polychlorinated biphenyls (PCBs) become increasingly concentrated as they move up a food chain (**biomagnification**) increasing their potential for harm. Biomagnification is why, for example, pregnant people are advised to limit the amount of fish they eat. This prevents the fetus from getting too high an exposure to toxins during critical developmental windows. Why fish and not burgers? Biomagnification.

Food chains, however, are just one part of any ecosystem. Yes, species eat one another but what happens when dinner's over? This one-way flow of calories from producers to the top of the food chain wouldn't last long if there wasn't also a way to cycle **nutrients** back into the system. Everything would get bound up in living organisms and the ecosystem would grind to a halt. To get the system to work in perpetuity it needs two things: death and decomposition. All living things die and when they die, **detritivores** and **decomposers** break down the complex tissues and molecules the living organisms were made of into simple chemical forms called nutrients (inorganic molecules). This is a critical step if the ecosystem is to continue to function. Remember, the producers use sunlight, yes, but they also need these simple nutrients from the water, soil, and air if they are going to grow. Nutrients also provide them with critical elements needed to make particular molecules (e.g. nitrogen for making proteins). Now, we can diagram the movement of matter in an ecosystem or, to be more precise, the cycling of matter around and around and around an ecosystem (Figure 5.3). One final point is that decomposition is just one way for matter to get back into the nonliving environment. Heaven forbid we forget about waste products! Remember species interacting with their environment included activities like poop, pee, barf, umm ... gas release, and so on. Burps happen. All of these activities also put matter and nutrients back into the nonliving environment (soil, water, air). Gross perhaps, but still true.

Looking at a diagram such as Figure 5.3 it is tempting to think that ecosystems are a perpetual life machine. After all, the matter (elements such as carbon, nitrogen, and oxygen) just keeps cycling around and around the system. This seems too good to be true and, it turns out, there is one other

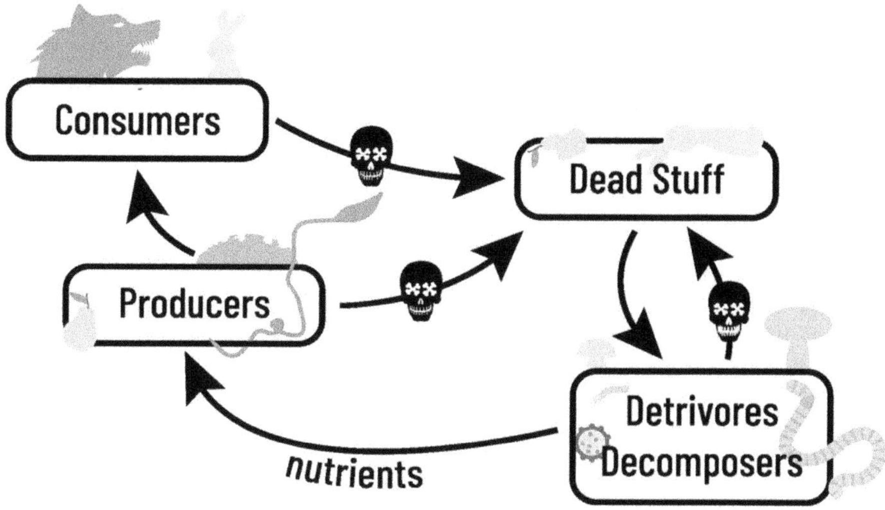

Figure 5.3 Matter cycles in all ecosystems
Note: Matter cycles in all ecosystems moving from producers to consumers and then, when individuals in these groups die, to decomposers and detritivores. Decomposers play the critical role of breaking down all the complicated macromolecules it takes to run a plant or animal. The breakdown product is those simple nutrients that the producers will then use to grow again.
Source: Original art by Professor Brytton Bjorngaard. Open Educational Resource.

critical nonliving thing that's not on earth that we need to keep things running. Yes, matter keeps cycling around – the journey of a single carbon atom for example is never ending. But the system cannot work without the constant input of **energy** from our beautiful, hot sun. Without the sun, the producers would not grow, they need to harvest the sun's energy (light) and, through **photosynthesis** convert it into a form of energy that consumers love: **chemical energy** in the form of sugar. The sun's light energy that is captured and converted does not cycle; it has a one-way flow through the ecosystem. Energy escapes quickly from ecosystems in the form of heat going back to the universe. This is the second law of thermodynamics in action. As chemical energy changes form, moving up the trophic levels, entropy increases, in this case releasing heat. Each step in the food chain is in fact very energy inefficient. For every ten calories at one trophic level, you only end up with about one calorie in the organisms at the next highest trophic level. If we think back to our burger example, that would mean for every 100 calories of grass a cow eats, it produces only 10 calories of burger for us to eat. The first law of thermodynamics tells us that energy has to be conserved and therefore has to go somewhere when an energy conversion takes place. Where does the energy that is *not* moving up the food chain (in this case 90 calories) go? This energy gets dissipated as a very dispersed form – heat – floating back out into the universe from whence it came. This heat generation happens as the ecological interactions and individual metabolisms of the plants, animals, and microbes involved play themselves out. To get back to our initial thought, in many ways our earth IS a perpetual life machine, but this machine needs the sun's energy to keep it running. We can think of the sun as life's battery.

Ecosystem Complexity and Indirect Effects

Although all ecosystems work in the ways I have discussed, there is astounding variation in the specifics of each particular ecosystem. This is both awe-inspiring and – from a practical perspective – incredibly frustrating. The frustrating piece is that this complexity means that we typically cannot predict how a system will respond to a given change until we explicitly study it. Even when scientists do study a given ecosystem a lot, they often find predicting changes difficult because complex systems behave in, well, complex ways. This can be frustrating for natural area managers and politicians who often want a quick fix and a simple answer. Why can't we just kill all individuals of a species if it is causing problems? Too often, quick fixes such as these come back to haunt us for multiple reasons. Later, we will consider pesticide resistance as one example when we look at examples of evolution. Right now, we will focus on the fact that every species in a system is embedded in what ecologists call an **interaction web**. Interactions webs include **food webs** – feeding relationships among species such as predator and prey – but they also include other types of relationships and interactions –species that benefit one another (**mutualists**), species that use similar resources for growth (**competitors**), and species that create habitat for other species. A beaver is an easy example of a species that creates habitat for other species – creating wetlands by damming creeks. Not surprisingly, these large, byzantine webs of interactions often exhibit very complicated behaviors. To make the dynamics of the system even trickier, and predicting outcomes tougher, some species interactions have strong effects and others weak effects. Ecologists therefore need to study any given ecosystem to measure interaction strength. Making things even more complicated, interaction strength can vary for a given pair of species across location. Have I mentioned ecosystems are complicated? One famous example of a strong species interaction was demonstrated by research on sea stars in the ocean's intertidal zone. Using removal experiments (rock faces where sea stars were removed), Robert Paine (1966) demonstrated that sea stars play a critical role in supporting the **biodiversity** in those rocky zones where the land meets the sea (Figure 5.4). Sea stars eat mussels and, when they are there, they eat enough mussels to keep the mussels from taking over the craggy rock surface. This allows prime real estate (open rock areas) to remain open for kelp (macroalgae commonly called seaweed), limpets, snails, and other marine invertebrates. This type of effect, where one species (in this case the sea star) influences a third species (the limpet) through its interaction with a second species (the mussel) is known as an **indirect effect**. Such indirect effects of species mean it is hard to know how the removal or introduction of a species will affect an ecosystem until we study the

Figure 5.4 Intertidal zone at Shelly beach, Cronulla, Australia
Source: Webb (2014).

interaction. Based on this work, Robert Paine coined the term **keystone species**. Not only does the sea star cause an indirect effect. It has a disproportionate effect on the whole ecosystem because of what it likes to eat. Finding such keystone species for systems is one of the jobs of an ecologist. Conserving these species to help the protect the whole ecosystem is the job of a conservationist.

Ecological Scales and Niches: From Populations to Ecosystems

Ecologists study ecosystems at different biological scales. One scientist might be interested in determining how we can increase the **population** size of a single, threatened species. Another might want to know whether a **non-native** plant drives **native** plants out of a community. A third might be interested in the cycling of nitrogen in a system. We would call these scientists population ecologists, community ecologists, and ecosystem ecologists, respectively. This is because of the biological scale at which they are most interested in measuring outcomes: changes in a population, interaction between two or more species, cycling of a nutrient. However, ecosystems remain the basic unit of nature regardless of which scale an ecologist is working at. This is because certain components are required in the nonliving environment for there to be any life. The exact nature of these components differs from species to species, but no species can survive in a vacuum, at least not forever. Interestingly, some species such as tardigrades, commonly called water bears, are *incredibly good* at surviving in extreme conditions including, in the case of these tiny creatures, outer space. Spores and seeds are other examples of living things that can lay dormant and alive until conditions are right for growth and reproduction. Even these species though need an ecosystem to complete their life cycle. They may be able to *survive* on just about nothing but they can't grow or reproduce. This means, like all species, they need an environment with the proper habitat to maintain their populations. Ecologists use the term **niche** requirements to refer to what must be present in an ecosystem if a given species is to live and flourish there. One way to think about niche requirements is that they are the conditions and resources that a particular species needs to be able to survive, grow, and reproduce. If these conditions are met, the site serves as habitat. Think of a fish; it will obviously not do well on land because it derives its oxygen from the water, along with many other adaptations for aquatic life. Humans have niches too – you would probably not do well in a deep-sea vent. So, a fish needs water, oxygen, bugs to eat, and several other environmental conditions. Depending on what type of fish it is it might require fresh or salt water, warm or cool water, and so on. If its **habitat** has these conditions and resources, that habitat is currently meeting the species' niche requirements. Niches are a good topic to bring us to our discussion of evolution. A species' niche requirements have a major genetic component; a fish's genetics determine for example how much oxygen it needs in the water. Environmental change is influencing species in certain cases by favoring some individuals within a given species over others. In the case of our hypothetical fish, IF (and that's a big if) there are individuals in the species that can do well with lower oxygen those individuals are more likely to survive and reproduce in a polluted, low oxygen lake. There will therefore be more individuals that can survive these lower oxygen conditions in the next generation. Hopefully, this example helps illustrate the tight coupling of ecology and evolution (environmental change and genetic change). This hypothetical case is adaptive evolution because the genes allowed individual fish to survive in lower oxygen leading to more individuals with this trait in the next generation.

How Evolution Works and Why It's (Sometimes) a Pain

I have defined evolution as genetic change over time. It's incredible to realize that Charles Darwin did not know about DNA/genetics and yet here I am stating that genes are the basic unit of evolution. Darwin was able to construct his theory of evolution just by observing that parents passed their traits on to their offspring. Now of course we understand that genes explain this heredity and therefore provide the mechanism for evolution to occur. This understanding is called the **modern synthesis**. Evolution has other definitions. It can also be defined as descent with modification which refers to the fact that species can change over time to whole new species! This, after all, is how you get birds from dinosaurs. Here, we are going to focus on one type of evolutionary change: adaptive evolution. Genes that code for traits that increase survival and reproduction will be favored by **natural selection**. Scientists call the traits associated with these genes, **adaptations**. Two major types of evolution are evolution that occurs within a species (**microevolution**), and long-term divergence of species into new taxa (**macroevolution**). To some extent it is just the amount of time (in generations) that distinguishes these two types of evolution but when we see small-scale changes in a species it can be hard to wrap our heads around the fact that these types of changes can create major taxonomic differences when given millions and billions of years. Take for example the evolution of whales. The closest relative to whales that is still alive today are hippos but the common ancestor that led to each of these species is no longer on earth. Adaptations for aquatic life for hippos and whales evolved separately as the lineages both have land-based ancestors (UC Berkeley). When we look at an evolutionary tree (phylogeny) that maps this history out it is hard to imagine how small genetic differences lead to such huge differences. In no small part, this is because our brains have a hard time thinking over 60-million-year time scales. Another example of the incredible amount of change that can occur over this type of time period is the example of birds. Yes, birds are the surviving dinosaurs but they sure as heck look a lot different than they did a few million years ago.[6] Except for baby egrets, baby egrets totally look like dinosaurs (Figure 5.5).

Figure 5.5 Great Egret chicks in a nest in Moro Bay, California (*Casmerodius albus*)
Source: Baird (2007).

To understand **adaptive evolution**, it is important to focus on the sequence of steps that allow adaptations to increase in populations over time. The first key fact to remember is that evolution *cannot* anticipate future environments. To the extent that species are well-adapted, they are adapted to *past* environmental conditions. In the past, their characteristics were ones that led to better survival and reproduction. This can sometimes mean that an adaptation is no longer adaptive, because the environment is constantly changing. A trait that increased survival and reproduction in a past environment may not do so in the current environment. Evolution *seems* simple, but the order that things take place makes all the difference, just like you and me, evolution doesn't know what's coming next. When someone talks or writes about evolution it's important that they avoid terms that suggest any anticipation of the future environment. With that warning in place, let us look at why, sometimes, evolution can be the cause of quite a few headaches.

As an example of the types of problems evolution can create, let's look at our desire to control populations of insect pests. It would initially seem as if the most effective way to control pests is to kill them: if you kill them they cannot eat the crop you are growing. We have become very good as a species at killing things, whether it is spraying glyphosate to kill plants or permethrin to kill insects. If you were a farmer, and saw insect damage to your sweetcorn crop you might decide to treat it with permethrin. The vast majority of insects that were damaging your corn would then be killed by the chemical and you would stop seeing damage, sell your corn, and hopefully have a good year. All seems well but it may not be, long term, because of evolution. In some cases, a small percentage of the insects you sprayed have genes that allowed them to survive despite the insecticide. These are the ones who, although small in number early on, survive the spraying. Now there is a lot of corn around and not a lot of **competition** because most of the other insects are dead so the surviving insects feast away, grow fat, and lay lots and lots of eggs. Remember, the unit of evolution is the gene. What genes do all those eggs have? They have the genes of the insects that can survive spraying. Evolution has happened. The genetic makeup of the insects has shifted in the next generation, now you have a population with a higher percentage of genes that code for **insecticide-resistance**.[7] What happens next? The farmer plants corn again the following year, they see insect damage so they spray permethrin because it worked so well before but unlike in years past, now very few of the insects die. Once again, the farmer has a problem. Actually, now they have two problems: crop damage and an insecticide-resistant pest. This is why adaptive evolution can be such a pain – it makes problems worse. The insecticide that used to work great, declines in its usefulness over time as the insect population adapts to the new environment (environment with insecticide). This is the exact same type of headache we have with medical drugs which can lose their power to treat infections over time. The antibiotic that was so powerful in past patients now has less effect because more bacterial strains have genes that confer resistance. In short, species can adapt to many of the methods we develop to eliminate them, making those methods less effective over time.

In its defense, evolution can also be a wonderful, beautiful, crazy-cool thing. (It turns out that people get to impose values – decide what is good or bad – values come from us not nature.[8]) As a conservationist, one of my favorite facts about evolution is that it is a tool in every species' tool-kit. Adaptive evolution gives them a fighting chance at surviving the major environmental changes we continually create. Evolution is by no means a silver bullet because species need to have individuals within their populations that do better in the new environment if they are to have a chance at adapting to changing conditions. Despite this constraint, we do find cases where evolution has allowed species to persist despite an environment we have dramatically altered. One example is in the world's coral reefs. As ocean waters heat up, coral that are more heat resilient are doing better. Some evolutionary biologists are even working on what is called **assisted evolution** – developing strains of species that should be extra resilient to new environments. One example of this is work being done by Kira Hughes and others at the Coral Resilience Lab in Hawaii; they are lab-raising corals that are tolerant of higher ocean temperatures[9] with the goal of introducing these "super corals" to our beleaguered oceans (NPR, 2021). Evolution is also incredible in that it is the process by which the phenomenal biodiversity of the earth has increased over billions of years – slowly creating the staggering variety of species we find today. A glance at diatom diversity can illustrate this point. Thanks, evolution (Figure 5.1)! It is evolution that gives us the viceroy butterfly that looks so much like a monarch butterfly (Figure 5.6). Why do these two species look so similar? Evolution. Monarchs store toxic plant chemicals in their bodies by eating milkweed. Birds therefore avoid eating monarchs because they taste nasty. No one likes a stomach ache. In

Figure 5.6 Is it a monarch butterfly?
Note: The answer is yes for the butterfly on the left but no for the butterfly on the right. The second butterfly is a viceroy, a species that gains protection by mimicking the monarch's coloration pattern
Source: Ramsey (2008).

this evolutionary environment, looking like a monarch butterfly (even if you are not one) is therefore advantageous. So, viceroys that looked like monarchs had higher survival as birds avoided eating them and ate the viceroys that looked more like ancestral viceroys. This type of evolution, where another species evolves to look like a venomous or toxic species occurs in many different evolutionary lineages throughout nature including, for example, king snakes that evolved to mimic venomous coral snakes. Evolution of such look-alikes is called mimicry.

Conclusion

To wrap up this chapter let's revisit the five principles we started with. Several of the things here hopefully reinforce the principle: "change is the one constant." So many changes are taking place continually within ecosystems. Material cycles. Energy has a one-way flow through the system. Species interact in ways that are not just feeding relationships. Each ecosystem is frustratingly and gorgeously complicated and unique. This complexity means that how we manage an ecosystem changes over time as we study these systems and find out more about how each one works and apply that new knowledge to new management approaches. Ecosystems are also changing because the species in them are constantly evolving (Box 5.1), the genes each generation carries are a little different from the previous generation. Plus, all these species are part of the environment for other species. There is a lot of change, all the time, and I didn't even talk about **ecological disturbances, succession**, or dispersal/migration. The living parts of the system are dependent on the nonliving parts. You need oxygen in your lungs and water in your cells. Every time you visit the restroom you are contributing matter to the environment. It is miraculous that with all these changes the earth stays viable for life and has the right environmental range for humans in particular. But, this could be changing. One tiny branch of the evolutionary tree, what some have called a "naked ape" that developed a big brain and opposable thumbs, is shifting the entire global environment. How significant is this shift and what does it mean for us and other species? We will take up that question in Chapter 7. Before that, let's spend a little bit more time thinking about those "endless forms most beautiful," what today biologists would call **biodiversity**.

Notes

1 Ecosystems have thresholds and when they reach them they can shift from one state to another. This is called a **regime shift**. Often several reinforcing factors keep them in one state until the threshold is met and then they move quickly to a different state. In the lake example discussed in this chapter, the likely culprit is nutrients (nitrogen and/or phosphorous) flowing into the lake and eventually forcing it to a new algal dominated ecosystem.

2 Dissolved organic carbon (DOC) refers to small carbon-containing biomolecules that are dissolved in water. It is a complex mix of various substances including well known molecules such as sugars or proteins. DOC is important for many reasons including serving as a nutrient for aquatic microbes.

3 If you are interested in finding out more about the work of art professor Brytton Bjorngaard who created Figures 5.2 and 5.3, check out her webpage at https://bryttonbjorngaard.com.

4 If you want to find out more about genetic drift (and who doesn't) you can watch a fun cartoon video from the Amoeba Sisters (2017) that explains it quite well.

5 It is hard to come up with a simple word to use for the planktonic producers that are at the base of aquatic food chains. Although they have chlorophyll and use photosynthesis to produce sugars, they are not plants taxonomically (they lack roots and stems). It is most accurate to call them phytoplankton but I will sometimes call them algae for simplicity's sake or single-celled algae given that macroalgae (such as kelp aka seaweed) is multicellular. Phytoplankton are a diverse group of single-celled species that conduct photosynthesis in aquatic systems.

6 I love the fact that birds are modern dinosaurs and could probably monologue about it for a couple more paragraphs. The goal though is to keep this book short and sweet so I will just say you should *read more* about this from articles like Luis Chiappe's open-access work *Downsized Dinosaurs* (Chiappe, 2009).

7 It is overly simplistic to think of a gene or even a group of genes as corresponding specifically with a given adaptation. Just like combining ingredients to make a cake a lot of genes work in complicated ways to produce traits. Please forgive my simplified language; I'm using it to make the process clear.

8 All logical fallacies are good to know but the appeal to nature fallacy is particularly relevant here. One version of this fallacy would be to claim that natural is equivalent to good, just, or inevitable (Logical Fallacies). This is a fallacy as there is no reason that something natural should be any better than something unnatural or cultural. Arsenic will kill you and is "all natural." Another example where natural isn't good comes from the field of animal behavior where we find male lions committing infanticide. John Cook's "Cranky Uncle" app also teaches about logical fallacies in the context of environmental science if you'd like to explore fallacies further.

9 People have different opinions on whether it is ethical for us to mess with coral evolution. Some argue we are ethically obligated to help species that are at risk from our activities. Others argue humans should just leave nature alone for once. What do you think?

References

Amoeba Sisters. (2017, June 8). *Genetic drift*. [Video]. YouTube. https://www.youtube.com/watch?v=W0TM4LQmoZY.

Atwood, M. (1985). *The Handmaid's Tale*. Toronto: McClelland & Stewart.

Baird, M. (2007). *Casmerodius albus4*. Retrieved from Wikimedia commons: https://commons.wikimedia.org/wiki/File:Casmerodius_albus4.jpg.

Castells-Nobau, A., Mayneris-Perxachs, J., and Fernández-Real, J. M. (2024). Unlocking the mind-gut connection: Impact of human microbiome on cognition. *Cell Host & Microbe*, 32(8), 1248–1263. doi:10.1016/j.chom.2024.07.019.

Graham, D. W. (2023) "Heraclitus", *The Stanford encyclopedia of philosophy* (E. N. Zalta and U. Nodelman, eds.). https://plato.stanford.edu/archives/win2023/entries/heraclitus/.

Chiappe, L. M. (2009). Downsized dinosaurs: The evolutionary transition to modern birds. *Evolution Education and Outreach* 2, 248–256. doi:10.1007/s12052-009-0133-4.

Haeckel, E. (1904). *Kunstformen der Natur*. Plate 85: Diatomea. Retrieved from Wikimedia Commons. https://commons.wikimedia.org/wiki/File:Haeckel_Diatomea.jpg.

Johnson, M. T. and Munshi-South, J. (2017). Evolution of life in urban environments. *Science*, 358(6363), eaam8327. doi:10.1126/science.aam8327.

Littleford-Colquhoun, B., Clemente, C. J., Whiting, M. J., Ortiz-Barrientos, D., and Frere, C. H. (2017). Archipelagos of the Anthropocene: rapid and extensive differentiation of native terrestrial vertebrates in a single metropolis. *Molecular Ecology*, 26(9), 2466–2481. doi:10.1111/mec.14042.

Logical Fallacies. (n.d.) *Appeal to Nature*. https://www.logicalfallacies.org/appeal-to-nature.html.

Natural World Facts. (2022, April 22). *Hydrothermal vents: Oases in the deep sea*. [Video]. YouTube. https://www.youtube.com/watch?v=ECBbAjoEHWI.

NPR (National Public Radio). (2021, December 9). Darwin in a lab: Coral evolution tweaked for global warming. https://www.npr.org/2021/12/09/1062608089/darwin-in-a-lab-coral-evolution-tweaked-for-global-warming.

Paine, R. T. (1966). Food web complexity and species diversity. *The American Naturalist, 100*(910), 65–75. doi:10.1086/282400.

Ramsey, D. (2008). Monarch Viceroy Mimicry Comparison. Wikimedia commons: https://commons.wikimedia.org/wiki/File:Monarch_Viceroy_Mimicry_Comparison.jpg.

Silverstein, S. (1964). *The giving tree*. HarperCollins.

U. C. Berkeley (n.d.). Understand Evolution: The evolution of whales. https://evolution.berkeley.edu/what-are-evograms/the-evolution-of-whales.

Webb, J. (2014). Cronulla Rock Pool, Shelley Beach, Cronulla NSW Australia. Wikimedia commons: https://commons.wikimedia.org/wiki/File:Cronulla_Rock_Pool,_Shelley_Beach,_Cronulla_NSW_Australia.JPG.

6 Biodiversity

The Intricate Web of Life

Biodiversity refers to the diversity of life across all biological scales: from variation at the molecular level (genes) on up to whole species, biological communities, and ecosystems. It might surprise you to learn that this is a fairly young word. I'll start this chapter discussing its recent origin. When we think about biodiversity, our minds naturally wander to the sad reality of present-day extinctions. How fast are we losing biodiversity? Here we'll look at how the current extinction rate compares to the losses experienced in earth's five mass extinction events. We'll find that the news is both good and bad. The good news is humans have not *yet* caused a sixth mass extinction. The bad news is the rate of biodiversity loss is at the pace of past mass extinction events. This suggests we are in a critical window of opportunity, where our work to stop extinctions may be able to prevent a mass extinction. This chapter will also talk about why biodiversity matters, even for people who hate nature. This is because ecosystems provide a wide range of services that support all aspects of the human enterprise. Unfortunately, we often take these services for granted until they are degraded. Diving into specific groups of plants and animals, we'll see that insects make up the bulk of animal diversity. Due to this, we'll look at this group in particular – examining global insect decline. Given the millions of species on our planet it isn't practical to approach conservation one species at a time. We'll see that one approach conservationists take is to use information on one group of species to help protect other groups. Specifically, this chapter has a case study discussing whether primates can serve as such an umbrella group to aid biodiversity conservation in China.

The word biodiversity is younger than me (!) and although I may seem quite old to you I am really not *that* old. In geologic time, I'm but a newborn babe. Kids might learn this word in grade school now, but when I was chasing lizards and pretending my dog was a wolf during my backyard adventures in the early 1970s, there was no such word as biodiversity. Plants, animals, fungi, protists, bacteria obviously existed. Scientists had been describing life for centuries as naturalists and then as ecologists, but it wasn't until the 1980s that this new word bubbled up out of academia. I'm old enough to remember when biodiversity got flagged as a typo by

DOI: 10.4324/9781003725404-6

software. What happened in the 1980s that demanded a new word be birthed? Biologists, returning to their field sites, started to see whole scale loss of the habitats they studied, and along with the habitats, the species that called those habitats home. How could they communicate this staggering loss, unique ecosystems disappearing right before their very eyes? Out of that struggle to give voice to the voiceless plants and animals, came the phrase biological diversity which was later shortened to a single word: **biodiversity** – a short and simple word that could hold all of life (Oxford English Dictionary, n.d.; Sarkar, 2021). I joke with my students that perhaps we should shorten it even more and just call it Bio-D. "More Bio-D, for you and for me." The chants write themselves. According to the Oxford English Dictionary, the first known use of the word biodiversity was in the 1980s in the academic journal *Bioscience*. I love this young, plucky upstart of a word; all of life on our marvelous planet crammed into 12 little letters. It can be defined in a few ways but I prefer a simple definition: the diversity of life across all biological scales. The three major biological scales are genetic, species, and ecosystems. **Genetic diversity**: all the myriad variations of genes within one species and among all the species of the world. **Species diversity**: the number and amount of all the different species. **Ecosystem diversity**: all the ecosystems from sea to shining savannah. One of the astonishing facts about evolution is that over the 3.5–4-billion-year history of life on earth, the planet has slowly but surely gained biodiversity. Despite five mass extinction events, biodiversity has bounced back each and every time and, if we ignore the last couple hundred years, we have more species now on our planet than at any previous time. Granted, "bouncing back" in geologic time takes *forever* in human generations – millions of years to recover a given level of biodiversity after it is lost, but nature is more patient than the most patient human, and evolution finds a way. **Speciation** happened again and again, as populations of a single species separated and the separate populations' genes shifted over time, creating genetic barriers so that the gene pools would never meet again.

If we step into an oak forest diverging from a nearby road, immediately the air feels cooler and the sounds from nearby traffic become muffled, maybe we hear scratches as something scurries up a nearby tree. A jay on a high branch might squawk at our intrusion as other birds sing in the background, desperate to find a date or to proclaim a spot as theirs and theirs alone. How much biodiversity are we going to need to build a functioning forest such as this? The first species we will need are the **dominant species** – those that are everywhere throughout the forest. Some of these dominant species may also be what biologists call **ecosystem engineers**, for example, the mighty oak. A mature oak tree can reach more than 100 feet in height. Its canopy leaves and branches towering stories above us. Even with just this single species on our list we can talk about diversity. Specifically, we can discuss **genetic diversity**. Every acorn, every grown tree, has a different genetic composition. Oaks have both male and female flowers. Yes, oaks have flowers they just aren't bouquet worthy. Oak tree flowers are very small and discrete; they are pollinated by wind which, unlike insects, doesn't need to be drawn to the plant by brash beauty and strong scents. Oaks outcross which means they need pollen from a different individual than themselves to develop their acorns. In a forest thick with oaks, the acorns on an individual tree would all have the same mother but they would have many different fathers depending on whose pollen reached any given flower on any given branch. So even within just one acorn-laden tree we have a wide diversity of genotypes due to parentage. Lots of variation in those tree babies (Figure 6.1)! Now cast your mind to all the trees in the forest, all the forests across the land, all the places we find this one species of oak throughout its geographic range. *That* is genetic diversity and it is the fodder for evolutionary change. If we want the species we love to persist, one of our best bets is to keep their population numbers soaring as high as we are able; with this abundance, we keep their associated genetic diversity high. Who knows what genes our oak will need in future environments? I don't. Evolution surely doesn't. Best to keep as many genes and versions of genes as possible just in case. Oaks comprise only a handful of species, albeit a critical group for the forest – the one that gives the forest its name.

Figure 6.1 Variation in size and color of acorns collected from one species (English oak, *Quercus robur*)
Source: Carter (2022).

Along with all the different species of oaks, there will be many additional species of plants, big and small: hickories, ash, maple, walnuts, hackberries, sassafras, paw-paws, spice bush, gooseberry, prickly ash, Virginia creeper, blue bells, snakeroot, violets, may-apples, green dragon – the list goes on and on. Animals, mainly insects and other arthropods, crawl, scurry, burrow, fly through the forest. Squirrels, of course, but also wasps, bees, grasshoppers, spiders and mites, isopods (aka roly-poly bugs) and earthworms. Mushrooms and other fungi are there including the mutualistic fungi that bring nutrients to the tree roots in exchange for sugar from the tree – the mycorrhizae. The soil has its own microbiome – teeming with microbes that do the hard work of crunching down big biomolecules such as lignin into the tiny nutrients a plant's roots need to grow (Figure 5.2). All of these species (and all the many more I haven't listed) are the next biological scale up from genetic diversity – they are the species diversity of the forest. But I've left one species out, that you all know incredibly well, *Homo sapiens*, us.

People have lived in and used forests for millennia. Why then is it so easy to exclude ourselves when we discuss the species diversity found in a forest? Part of the answer goes back to the discussion from Chapter 4; western cultures banish humans from what is natural – placing us outside of their concept of nature. Often this banishment from nature is coupled with viewing nature's goods as commodities – nature is the place where resources are extracted and waste deposited. It is over there. People though are very much part of natural systems, embedded and dependent upon a healthy functioning planet and local functioning ecosystems. One of our introductory eight principles is just that – you are part of nature. I'll add to that – whether you like it or not. From the air we breathe, to the crops we grow, to the wooden tables we sit at, we are of and in an ecosystem. Which brings us to our next scale of biodiversity: ecosystem diversity. We'll end back in our oak forest, on a nature trail, but let's start our next ecosystem tour in a place you might not expect to find nature – in the heart of a bustling urban downtown.

Cities have been called concrete jungles and given the degree of built infrastructure in a thriving city it can take a minute to see the urban ecosystem – but it is there. There may be a concrete walkway next to a river running through town but most of the streams that once braided themselves through the landscape have been forced underground with water running into the gutters and down into the network of sewer systems. City soil is covered primarily by **impervious surfaces**, cement and asphalt, that don't absorb water. In many locations, the only green one sees is a monoculture of lawn grass. The urban park – with its forest or meadows or desert – grows in a sea of asphalt. Along the road edges, urban trees may reach for the sky among the buildings. Here and there, more biodiverse plantings or parks occur, a roof-top garden, a patch of pollinator plants, a city park. Although ecologists would not call this a jungle, we would call it an ecosystem. Living beings interacting with the nonliving environment. Ecologists call cities urban ecosystems. There is a whole sub-field of ecology, urban ecology, which studies a wide variety of topics such as nitrogen cycling in urban lawns. Another highly human-dominated ecosystem would be an agroecosystem, more commonly called a farm. Here, again, we can have monocultures, but now instead of lawn we have monocultures of crops interacting with the soil and the air, and in some farm fields prairie strips that follow the topographic ridges across the farm and help boost biodiversity, soil, and nutrient retention in the field. Natural vegetation strips in cropland are a wonderful way to improve environmental quality. Researchers at Iowa State University looked at the impact of putting 10% of farm fields into strips of prairie.[1] They found dramatic improvements in soil and nutrient retention along with increases in plant and bird biodiversity (Iowa State University, n.d.).

Ecosystems are everywhere so as we walk from our urban ecosystem, through an agroecosystem, we might see other "more natural" ecosystems. Taking a trail into a local natural area we might find a stream, a lake, grassland, forests. Depending on where we find ourselves on the planet, we might walk to a dune or the seashore, dip our toes in the vast marine ecosystem we call an ocean. Marvel at the variety of plants and animals we find as we explore this last scale of biodiversity: ecosystem diversity. Maybe we are near an ocean with rocky tidal zones which contain sea stars, barnacles, snails, small fish, anemones and much more. And as we sit with our feet in the sand, we remember that ecosystems can also be very tiny. Microbial communities on our skin, in our guts, make us and other creatures walking, swimming, flying ecosystems (Figure 5.2).

So, there we have it. One little word, biodiversity, that holds a planet teeming with life. An incredible fact about biodiversity is that we haven't even cataloged it all – given each piece a name and description – let alone studied how all these myriad species live and interact in their habitats and interaction webs. Scientists cannot even tell you how many species there are on earth. There are estimates, of course, and better educated guesses for some taxonomic groups than others. In 2011, Camillo Mora and others used a new technique that estimated there are between 7.4 and 10 million eukaryotic species on earth. This estimate leaves out bacteria and other prokaryotes for a number of fairly academic reasons;[2] if they were included the estimated number of species would be much higher. We know for certain though that we share our planet with millions of species, most of which we know very little about. This is one of the joys of being a biologist. There is new biodiversity to discover and study and not just in remote areas. In 2024, thousands of new species were discovered. Some, such as four species of deep-sea octopi, were found by exploring areas of the sea little traveled. But others, including the vampire hedgehog, were lying in dusty museum drawers. It took genetic comparisons of preserved animals from multiple museums to find that little vampire (Adkins, 2024). A vampire hiding in a museum; it seems the stuff of fantasy. As exciting as these discoveries are they can be bittersweet when we view them in the context of biodiversity loss and species extinctions. A sad truth is that untold numbers of species will go extinct before we even get a chance to peer at them through a microscope, or watch them wiggle in a net, before we can even do the bare minimum – describe them and give them a name.

Everything You Ever Wanted to Know About Mass Extinctions

Humans are now what Paul and Anne Ehrlich (2008) called the "dominant animal" on the planet – and with our dominance has come global level environmental change. One of these changes is catastrophic biodiversity loss. This loss is occurring at all levels of biodiversity: loss of genes and genotypes, species extinctions, and whole scale ecosystem loss. Small populations of species by default contain fewer genes than large populations so as populations of plants and animals dwindle in size, they lose the very thing that helps them do well in a changing environment – genetic diversity. Sometimes whole ecosystems are threatened by the human enterprise. The grasslands of North America – tallgrass and shortgrass prairies – are one such ecosystem, classified as a critically endangered biome because their loss greatly exceeds their level of protection (Hoekstra et al., 2008). The deep roots of prairie plants created some of the best soils on the planet and so grasslands including North America's prairies are imperiled globally. Across the planet grassland ecosystems have disappeared converted to low diversity agroecosystems (farm land). As human demand for land and other resources has increased, biodiversity loss has accelerated. Historically, people have converted habitat that is serving as home for many other species into habitat for one species – us – with little thought to the consequences. Right now, at the beginning of the 21st century, we live at a critical time, we have a window of opportunity that is closing. The numbers are stark, we are losing species at a faster rate now than in the past five mass extinctions, but, the good news is, most of that biodiversity is still here. But with every passing day, the window of opportunity gets smaller as more loss occurs. It's time to wedge open that window and start saving biodiversity with abandon anywhere and everywhere we can.

The earth is approximately 4.5 billion years old. At first when the earth was young, there is no evidence of life. The estimates of when life first arose vary from 4.3–3.5 billion years ago depending on method. More than 3 billion years ago we start to see the prokaryotes – those simple single-celled species. They are the ancestors of us all, including being ancestors to the modern prokaryotes such as bacteria and archaea. Prokaryotes keep it simple having the most basic of cell structures – they lack the complex organelles, such as mitochondria or chloroplasts, that we find in eukaryotes. They have cytoplasm, a cell membrane, DNA, and proteins such as ribosomes. These simple forms of life have all the basics a cell needs to keep making other cells but not much more. No frills. No bells or whistles. They take a simple, unsophisticated approach to life. Pack light, just the necessities. In contrast, eukaryotes bring much more on their journey through life. These species have more complex cellular structures and organelles – they first appeared around 2.7 billion years ago. In the present day, there are millions of species of eukaryotes – the estimated total is between 7.4 and 10 million (Mora et al., 2011). Eukaryotes – including plants, animals, and fungi – arose over time with five, well, let's just call them major setbacks. In the history of the earth, there have been five mass extinction events where the world lost a staggering percentage of its species. Technically, more than 75% of existing species have to go extinct for an extinction event to qualify as a **mass extinction event** (Barnosky et al., 2011). This has happened five times in the history of life. A question I get asked a lot is whether we are currently in a 6th mass extinction event caused by human dominance. How would we know if we were? It turns out this is a harder question to answer than you might anticipate.

Biologists are fond of saying "the fate of all species is extinction." Of the estimated 4 billion species that have slithered across the planet at some point, 99% are no longer here. Extinction happens and over geologic time scales, which take practically *forever*, it happens a lot. So, extinction in and of itself, is less of a concern than the *rate* of extinction. How fast are we currently losing species? Specifically, are we losing them at the same rate that has occurred throughout the fossil record (background extinction rate) or are we losing them much faster, on pace with the rates seen only during the mass extinction events? Barnosky and others (2011) investigated this very question. They measured how fast species went extinct during the big five mass extinctions and compared that rate to how fast we are currently losing species. What they found is humbling. We are currently on pace for a mass extinction event. The current rate of extinction is much faster, 100 to 1000 times as fast as the background rate. In fact, the current rate of extinction is even *faster* than the extinction rates that resulted in the big five mass extinction events. That's the truly devastating news. There is, however, some good news to temper the bad. The good news is, for the most part, the incredible species we share the planet with are still here. Although we are losing species incredibly rapidly, we are nowhere near the 75% loss threshold. What percent have we lost? For many understudied groups, the reality is we just don't know. But we do know the answer for well-studied groups such as our closest kin – the other mammals. For mammals, around 25% are either already extinct or threatened with extinction (IUCN, 2008). Threatened species are combined with known extinctions when estimating rates because of the difficulty of proving extinctions.[3] How long would it take the current extinctions to reach a mass extinction level loss? If all threatened species go extinct in the next century and that rate of loss continues, we would see a 6th mass extinction in 240–500 years. In geologic time this is a blink of an eye, but for humans it is multiple lifetimes, meaning there is time to reverse course and start to think outside the box about where we can help other species thrive. It will be up to the next few generations to save as much biodiversity as we possibly can. Barnosky and others (2011, p. 10) sum it up well when they write, "It is encouraging that there is still much of the world's biodiversity left to save, but daunting that doing so will require the reversal of many dire and escalating threats." The threats are real and climate change is just one of many.

With millions of species to protect, we don't have time to protect one species at a time. Luckily, there are more efficient ways of preserving biodiversity, including using an **umbrella species** approach (Box 6.1).

Box 6.1 Will protecting primates protect much more? Case study from Yunnan Province, China

With millions of species on earth, figuring out how to best approach biodiversity conservation is a challenge. It would be lovely if we had the resources to learn about each and every species, develop a plan for its conservation, and then implement that plan. This however just isn't realistic given the short time window we face in protecting the earth's diversity from very real threats. One technique that has been suggested as a work around for this problem is to find **umbrella species** and design plans for their conservation. Umbrella species are those whose protection in the wild will by default protect many other species. Typically, large-bodied vertebrates are suggested as candidates for this approach. This is because large-bodied animals need large areas to support their populations. Large areas should, the thinking goes, also protect lots of smaller plants and animals. Yang and others (2025) decided to use extensive datasets to determine if protection of non-human primates in the Yunnan Province of China could have an umbrella effect not just in terms of protection of other species but also in terms of preserving ecosystem services. They evaluated primates as a group – 16 species total – including gibbons, snub-nosed monkeys, langurs, macaques, and lorises. They found strong evidence that using primates as umbrella taxa would aid conservation of much of Yunnan's flora and fauna and would also help protect key ecosystem services. Areas with high numbers of primate species were also rich in seed plants, birds, reptiles, amphibians, and other mammals. In addition, these areas had high levels of carbon storage and high water and soil retention. This suggests that by protecting its primate species, Yunnan can protect its overall biodiversity and maintain critical **ecosystem services**.

Biodiversity Matters: Ecosystem Services

Some people, like me, are crazy in love with biodiversity. When I was a little kid, I loved chasing lizards, digging in the mud, playing with isopods, you name it. In all honesty I still love turning over a rock to see what I can find underneath it. As a professor, it's a thrill to take students into local streams and hear them scream and laugh when they catch a huge crayfish. I realize, however, that these types of activities aren't everyone's cup of tea, and biology students are definitely a biased sample of the world's people. Biased towards a love of the slimy and slithery. If you aren't one of the people who loves plants and animals, should you care? Does the fast rate of current extinctions really matter to you? The short answer is yes it does. Even if you stay put in your concrete, urban ecosystem – losing species affects us all and sometimes in surprising and unexpected ways.

The origin story for the phrase **ecosystem services** is interconnected to the birth of the word biodiversity. This young phrase was also birthed in the early 1980s, scientists needed ways to communicate not just the staggering loss of unique life forms (biodiversity loss) but also the benefits we risked losing as this biodiversity declined (now called ecosystem services). Ecosystem services are all the goods and services ecosystems provide to us. To some extent it is a very hubristic concept, putting people at the center of it all. What has nature done for ME lately? The answer is a heck of a lot more than most of us realize. Scientists classify ecosystem services into four categories: supporting services, provisioning services, regulating services, and cultural services. **Provisioning services** are the ones we tend to know about. Yes, apples come from apple trees. These services are when nature provides us with, well, provisions. We tend to understand that we need to mine certain metals to build our technologies. We do after all need oxygen to stay alive so it is good that the phytoplankton in the ocean keep pumping it out. Yes, I said what I said, the phytoplankton. Trees tend to get credit for this work but the truth is it is the small primary producers, the single-celled plant-like algae that deserve the credit for the oxygen we breathe with about 50% of the atmosphere's oxygen coming from our oceans (NOAA, 2024). Have you thanked a *Prochlorococcus* today?[4] The list of goods provided by nature goes on and on – wood and food from a local forest, fruits, vegetables, grains from agroecosystems. Fresh water from lakes and streams. Provisioning services are just the most obvious of the ecosystem services – a direct material benefit of another species' existence is clear (e.g. an oak tree to make into an oak table). Some provisioning services are less well known – for example around half of all medicines are originally derived from a natural source (Wong, 2001). But there are millions of species on the planet and lots of them are insects, do we really need *all* these species? What's a parasitic wasp done for me lately? Parasitic wasps bring us, naturally, to the next category of ecosystem services: **regulating services**.

Regulating services refer to the role ecosystems play in regulating the state of our environment: the climate, the water cycle, and levels of pests and disease. For the water cycle, not surprisingly, we prefer to have clean water and to also not have our cities flood or our crops become parched. Just for fun though let's start with a regulating service that is more obscure than the water cycle. Everybody likes a nice cool glass of clean water. Less loved is my friend, and hopefully soon to be your friend, the parasitic wasp. Parasitic wasps parasitize other insects. They do this in a particularly ghoulish manner. The mom grabs a prey, say a caterpillar, and lays her eggs *inside* it without killing it. The caterpillar crawls away to keep eating and pumping food into itself to feed the now growing wasp eggs. After a period, these eggs hatch and the wasp larvae, eat their way out of the caterpillar killing it in the process. Basically, it's like the movie *Alien* but it happens right here every day on earth. How in God's name could this be a good thing? It just so happens that a number of the prey species for parasitic wasps are pest insects. By keeping these other insect populations in check, wasps indirectly help people's crops grow without insect damage. Lots of wasps, no pest outbreak, yummy food for us. Hopefully, you now see the benefit of having parasitic wasps in the world. Insects as a group, like other groups, are in decline. How severe a decline? I look at this in Box 6.2.

Box 6.2 Insect apocalypse: How severe are insect declines?

When some people hear the phrase insect apocalypse they no doubt think of a swarm of locusts or some other horrific B-movie scene involving insects, cockroaches maybe, in exuberant abundance. Ecologists however associate this phrase with a low abundance of insects and insect decline as well as with a debate within our field about the degree and nature of this decline. Insects are a hyper-diverse group of animals – making up the bulk of described animal species. They are understudied in comparison to vertebrates (mammals, reptiles (birds), amphibians) and range from the beautiful and essential to the pest-like and annoying. How are they doing as a group? Given the decline in vertebrates it would be foolish to think insects were doing better. Many of us have lived long enough to remember when a higher abundance of land-based insects meant insect-splattered car windshields that needed multiple cleanings on any road trip. I remember cleaning a lot of insect parts off of our cars when my dad would pay me to wash them back in the 1970s. These cultural memories make it pretty clear there has been a decline in land insects but how severe is their decline and which groups in particular are at risk?

The phrase "insect apocalypse" comes from a 2017 article that compared insect biomass at sites in Germany over time – 1989 to 2016 – and found a 75% decline (Hallmann et al., 2017). A paper published the following year (Lister and Garcia, 2018) showed even larger insect declines over the same period in Puerto Rico. These papers caused a bit of a panic among ecologists and conservationists with a number of subsequent articles using dramatic titles with phrases such as "insectgeddon" or "insect apocalypse." Not surprisingly, these articles caught the eye of the popular press and opinion sections. One outcome to emerge from this attention was a much-needed surge in studies studying insects and increased public awareness of insect conservation. As often happens when a phenomenon is better studied, the picture painted became more nuanced and complex with added scrutiny. Insects are a diverse group and some groups of insects showed positive trends over time, while others were in major decline. **Specialists** were more likely to be declining compared to **generalists**. Insects that are further up the food chain were more at risk. Aquatic invertebrates (including insects) seemed to be doing fine at least in some countries. We really don't know how insects are doing in critical regions such as the tropics because ecosystems there are understudied compared to the U.S. or Europe (Blüthgen et al., 2023). What might be driving declines? Insects are **ectotherms** so their response to changing temperature is different than for **endotherms**. Some entomologists point to climate change as a driver of decline but even greater drivers seem to be land-use change (habitat loss) and agricultural intensification (insecticides, herbicides). There still is much to learn about the threats and the specific types of insects that are at risk. The scientific community needs to build research strength in understudied ecosystems by increasing educational access in countries of the Global South. Because insects are the major food source for most birds, their fate is undoubtedly tied to the fate of other groups. What we do know for certain is that insect decline is real and has consequences for the ecosystems we know and rely on.

Unfortunately, because regulating services work in these types of indirect ways, people tend to take them for granted until the service declines. Even when a service declines, if a system is poorly studied we may not understand why we've seen the decline (e.g. what caused the pest outbreak). Regulation of diseases and pests is just one example of a regulating service. Grasslands, forests, and our oceans all absorb literally tons of atmospheric carbon thus decreasing carbon dioxide which would otherwise heat our planet. They regulate climate at a local level as well. Neighborhoods with more trees are cooler in the summer than neighborhoods with few trees. Ecosystems also prevent flooding and suck toxins out of water making it cleaner. In one incredible study, ecosystems appear to have saved people's very lives. Saudamini Das and others (2009) found that the lives lost in Indian villages during a cyclone were directly related to the coverage of mangroves protecting each village: more mangroves, fewer lives lost. This puts a whole new slant on tree hugging! Soils in ecosystems also absorb water during storms (in contrast to concrete or asphalt which are impervious surfaces). This water absorption helps both clean the water and prevents it from getting to river systems too quickly, thus decreasing the likelihood the river will have severe flooding. It also helps to recharge local aquifers (underground lakes that serve as water sources for wells). Not to mention that water pulled up through a tree's root system isn't dripping into your basement. If my basement isn't flooding, maybe I have time to catch some fireflies in my front yard. Awe and wonder in a mason jar brings us to the next set of ecosystem services: cultural.

Cultural services are the psychological, recreational, spiritual, and educational benefits we get from nature. Some of my fellow biology professors study phylogenetics – trying to piece together which species are most closely related and building maps of these relationships called phylogenies or evolutionary trees.

Nature provides us with this intellectual puzzle to solve. Nature also has tremendous positive psychological effects (Nature Rx). Taking time in nature, for example a stroll in a nearby green space, can boost mood as well as improve memory and attention. Researchers have done controlled studies where some people walk through a heavily concreted area and others walk in a green space for the same amount of time. You can quantify the psychological benefit (Weir, 2020; read more about this effect in the final chapter, Box 9.1)! There is also the awe and wonder many people find in nature. For some, this is spiritual. The monarch butterfly, and its continentally long migration, is a key component of fall festivals in Mexico. Indigenous groups give great significance to the return of these butterflies each fall. In early folklore, the monarchs are returning souls of those who have passed thus linking the living to the dead during the festivals of *Dias de Las Mortes* (Day of the Dead; Merotto, 2022). Metamorphosis is a particularly lovely metaphor of transformational change, so this is but one of many examples where we find it woven into stories cross-culturally.

As indirect as the benefits of many cultural and regulating services are, they are still more direct than the supporting services. **Supporting services** are basic processes that occur in ecosystems that make all of the previously discussed services possible. You can't have crops without soil. You can't have regulation of the water cycle, without, well the water cycle. Ecologists would call these ecosystem functions. Decomposition for example is an ecosystem function. Pollination is another. Cycling of carbon, nitrogen, oxygen, phosphorous and other elements around and around different parts of an ecosystem. From the air to the soil to the living things and back again. The earth works in particular ways. We've already discussed how matter cycles in ecosystems and how the light energy from the sun is converted to chemical energy and then has a one-way flow through earth's ecosystems. All of these basic processes are necessary for life on earth including, of course, human life. As an ecologist I admit that this is where I become skeptical when a science-fiction movie talks glibly about terra-forming a new planet and changing the atmosphere to suite human life. I for one will not believe it until I see it. For now, I think we should focus on protecting the conditions on *this planet* that make it suitable for 7 million + species. We evolved on this planet and it suits our needs quite well.

I once asked my introductory class what they needed to survive. They got stuck on money for a long time. They really needed more money. I made my scenario more restrictive. Okay, you can't buy anything but I can bring things to you including cash, what do you need. You are now sitting in a room for hours in a pile of money which I keep bringing to you because you keep wanting more. At some point it became clear to them that food and beverage delivery would be nice. Are they allowed to leave and go to the bathroom? Food, water, shelter, clean air to breathe, a private and clean spot to put matter back into the nonliving environment. We are animals and at the most basic level we need the things that any mammal would need. But we live in a world where we are removed from contemplating these basic truths. Our vegetables come plastic wrapped from the grocery store. We breathe without even thinking. When we need the mammalian basics, we buy them. With that in mind, there is one last point to be made regarding ecosystem services. Our economy is dependent upon them and not just slightly. A report from 2020 from the World Economic Forum estimates that more than half of global gross domestic product (GDP) is dependent on healthy ecosystems (Russo, 2020). If we want a healthy economy, we need a healthy planet.

Conclusion

Biodiversity is threatened and with it the ecosystem services it provides. If I asked you to name the biggest threat to biodiversity, I would guess you would say the climate crisis. News about the climate crisis is everywhere and it no doubt *is* a threat to the world's biodiversity. However, it is just one of many major threats. Habitat loss has caused more biodiversity decline than our changing climate. We are in a new era where humans have altered the globe in historically unprecedented ways. Yes, we are changing the global climate but we are also polluting the land and waters, shifting natural forests and grasslands to plantations and fields, and more. In our search for a sustainable future, we have to understand the nature of a variety of threats. Often there are interactions or tradeoffs among threats. Our solutions will need to be creative - simultaneously addressing as many threats as possible. Global environmental crises are **wicked problems**; such problems are systemic, complex, interconnected, value-laden, and defy easy solution. The climate crisis is very real but so also is biodiversity loss, plastic pollution, and more. It is critical that these problems be addressed holistically with the aim of shifting entire systems along with problematic system outcomes. What exactly are these major threats? We'll look at six pressing global challenges in the next chapter. It's not just climate, friends.

Notes

1 Because I live in the "prairie state" I tend to use this word for North American grasslands forgetting that it is only used in this region. Regardless of the global location of farmland, adding natural vegetation strips to fields should enhance environmental quality.

2 Species is a tricky concept to define and species concepts are much debated. Depending on the species concept you're using you might get more or fewer species. For example, the phylogenetic species concept frequently can reveal cryptic species that we only find when we analyze DNA. In the study used here to estimate the number of species on our planet, all the prokaryotes are left out of the estimate (Mora et al., 2011). There are a lot of reasons for this. Bottomline – life is different if you're a bacterium! For instance, horizontal gene exchange – picking up other DNA from distantly related bacteria – can happen. This makes species delineation for this group tough.

3 It turns out that verifying an extinction is incredibly hard. How does one go about showing that there are absolutely no individuals of a species left on the entire planet? Has sufficient effort been spent searching? Because of this challenge, what tends to happen is scientists just stop seeing individuals over time and slowly the truth of the matter seems evident as the "last seen date" moves farther and farther into the past. Once in a while, however the **Lazurus effect** is seen where a species we all thought was gone forever happily surprises us by popping up where we very much did not expect it.

4 If you want to thank an ecosystem for the oxygen you breathe, thank the sea. If you want to thank a particular group of species, thank bacteria in the genus *Prochlorococcus* the smallest photosynthetic organism on earth (NOAA, 2024). The Amazon forest is an incredible carbon sink and home to thousands of species – a biodiversity hotspot – but it is not the "lungs of the planet" (Zimmer, 2025).

References

Adkins, F. (2024, December 29). Vampire hedgehogs, pirate spiders, and fishy fungi – the strangest new species of 2024. BBC. https://www.bbc.com/future/article/20241224-the-new-plant-and-animal-species-discovered-in-2024.

Barnosky, A. D., Matzke, N., Tomiya, S., Wogan, G. O., Swartz, B., Quental, T. B. ... and Ferrer, E. A. (2011). Has the Earth's sixth mass extinction already arrived? *Nature*, 471(7336), 51–57. doi:10.1038/nature09678.

Blüthgen, N., Dicks, L. V., Forister, M. L., Outhwaite, C. L., and Slade, E. M. (2023). Insect declines in the Anthropocene. *Nature Reviews Earth & Environment*, 4(10), 683–686. doi:10.1146/annurev-ento-011019-025151.

Carter, W. (2022). *Acorns from a Quercus robur* (English oak). Wikimedia Commons. https://commons.wikimedia.org/wiki/File:Quercus_robur_acorns_in_Tuntorp_1.jpg.

Das, S. and Vincent, J. R. (2009). Mangroves protected villages and reduced death toll during Indian super cyclone. *Proceedings of the National Academy of Sciences*, 106(18), 7357–7360. doi:10.1073/pnas.0810440106.

Ehrlich, P. R. and Ehrlich, A. H. (2008). *The dominant animal: Human evolution and the environment*. Island Press.

Hallmann, C. A., Sorg, M., Jongejans, E., Siepel, H., Hofland, N., Schwan, H. ... and De Kroon, H. (2017). More than 75 percent decline over 27 years in total flying insect biomass in protected areas. *PLOS One*, 12(10), e0185809. doi:10.1371/journal.pone.0185809.

Hoekstra, J. M., Boucher, T. M., Ricketts, T. H., and Roberts, C. (2005). Confronting a biome crisis: global disparities of habitat loss and protection. *Ecology letters*, 8(1), 23–29. doi:10.1111/j.1461-0248.2004.00686.x.

Iowa State University. (n.d.) *Science-based trials of row-crops integrated with prairies strips*. https://www.nrem.iastate.edu/research/STRIPS/content/about-strips.

IUCN. (2008). *IUCN red list reveals world's mammals in crisis*. https://iucn.org/content/iucn-red-list-reveals-worlds-mammals-crisis.

Lister, B. C. and Garcia, A. (2018). Climate-driven declines in arthropod abundance restructure a rainforest food web. *Proceedings of the National Academy of Sciences*, 115(44), E10397–E10406. doi:10.1073/pnas.1722477115.

Merotto, T. (2022, October 31). Winged messengers: How monarch butterflies connect culture to conservation in Mexico. *Smithsonian magazine*. https://folklife.si.edu/magazine/monarch-butterflies-mexico-culture-conservation.

Mora C., Tittensor D. P., Adl, S., Simpson, A., and Worm, B. (2011). How many species are there on earth and in the ocean? *PLoS Biology* 9 (8): e1001127. doi:10.1371/journal.pbio.1001127.

NOAA. (2024). *How much oxygen comes from the ocean?*https://oceanservice.noaa.gov/facts/ocean-oxygen.html.

Oxford English Dictionary. (n.d.). *Biodiversity*. https://www.oed.com/dictionary/biodiversity_n?tab=factsheet&tl=true#20265023.

Sarkar, S. (2021). Origin of the Term Biodiversity, *BioScience*, 71 (9), 893, doi:10.1093/biosci/biab071.

Russo, A. (2020, January 19). Half of world's GDP moderately or highly dependent on nature, says new report. *World Economic Forum*. https://www.weforum.org/press/2020/01/half-of-world-s-gdp-moderately-or-highly-dependent-on-nature-says-new-report/.

Weir, K. (2020). *Nurtured by nature*. American Psychological Association. https://www.apa.org/monitor/2020/04/nurtured-nature.

Wong, K. (2001, April 9). Mother nature's medicine cabinet. *Scientific American*. https://www.scientificamerican.com/article/mother-natures-medicine-c/.

Yang, Y., Li, C., Fang, Y. H., Ren, G. P., Huang, Z. P., Li, Y. P. ... and Xiao, W. (2025). Effect of primate protection on threatened and endemic vertebrates, plants, ecosystem services, and future climate refugia. *Conservation Biology*, 39(5), e70019. https://doi.org/10.1111/cobi.70019.

Zimmer, K. (2025, August 19). Why the Amazon forest doesn't really produce 20% of the world's oxygen. *National Geographic*. https://www.nationalgeographic.com/environment/article/why-amazon-doesnt-produce-20-percent-worlds-oxygen.

7 Global Change

A Planet Out of Balance

Humans have rapidly altered the global environment in novel ways. For example, the earth has been in a relatively stable climatic period for the last 10,000 years but that is changing. Climate change, however, is just one of a suite of environmental changes which are referred to collectively as global change. These include but are not limited to: ocean acidification, invasive species, nitrogen and phosphorous loading, atmospheric change, land use change, biodiversity loss, and novel entities such as plastics. This chapter examines these changes in the context of planetary boundary research – which seeks to determine if these various global changes are pushing our planet past stable conditions. Dauntingly, this research finds that six of nine planetary boundaries have now been crossed putting the earth outside its safe, Holocene-like, zone. The planet, of course, will be fine but what about people? Because a wide variety of issues place the planet at risk, solutions to environmental problems must be holistic; they need to be implemented in a manner that avoids trading one environmental problem for another. Two case studies in this chapter examine the question of an Amazon forest tipping point and the challenge of plastic pollution.

Welcome to the Anthropocene: Wicked Problems

I first encountered the phrase **wicked problem** in a textbook focused on conserving global biodiversity. As you might imagine, stopping the rate of global extinctions is no easy task. The largest threat to biodiversity on land is habitat loss, and biologists don't have a magic wand we can wave to increase habitat for threatened species. Habitat loss is a result of the human enterprise in all of its messiness. People tend to take over areas when we are present and modify them to meet their own wants and needs. So, to address biodiversity loss you need laws, regulations, you need to address social norms. Quickly you leave the field of biology behind and enter the fields of law, policy, sociology, cultural anthropology, and psychology. The term wicked problem was in fact coined by two social scientists (Rittel and Webber, 1973) but is now broadly used across disciplines including reference to the seemingly intractable environmental challenges we face from

DOI: 10.4324/9781003725404-7

plastic pollution to the climate crisis. Balint (2011) describes problems as moving in difficulty from simple to complex and finally to wicked with this final stage characterized by a "high degree of scientific uncertainty and a deep disagreement on values." Gordian knots. I would add that because these problems are systemic – solutions to one part of a wicked problem also tend to have unexpected consequences, ripple effects that lead to new emergent problems or exacerbate existing issues. Attempting to solve wicked problems can be a bit like a game of whack-a-mole. Environmental challenges are wicked in nature because they involve multifaceted, interacting natural *and* social systems. A student in a discussion section once compared this to a fitted sheet: you get one corner of the fitted sheet on and the other darn corner pops off. We've already discussed in Chapter 5 why the natural part – ecosystem dynamics – are so difficult to predict. Because of the **interaction webs** in systems coupled with evolution; we can't just kill one unwanted species and move on to the next problem. Loss or gain of even just one species can have reverberating effects. Now couple that ecological complexity, with societal complexity and things became quite wicked, quite quickly. This fact has major implications in how we combat environmental challenges. The **precautionary principle** dictates we move slowly in our interventions given uncertainty in outcomes. Systemic, knotty problems also need to be studied holistically with an eye to determining how the parts influence one another. We also need to expect and accept imperfect solutions, at best we keep moving in the right direction, learning as we go. Our actions make things a bit better and, if we continue to gather information, and aren't ashamed to admit to any failures we made along the way, we keep learning. This knowledge then helps the next iteration of solutions be a bit better still. Wicked problems are not solved in a single generation or by a magic bullet (or wand). As I think more and more of us are realizing, as we experience climate shocks such as the 2025 Los Angeles fires, we need, to a large extent, to change everything. Business as usual is just not working.

Humans have significantly altered global ecology. We have built all of our civilizations of the past 10,000 years during the stable Holocene period but now we are pushing the planet into a new state. The scale and suite of environmental challenges are unique to our time. In ecology, we call such planet-scale alterations **global change** and we will do a whirl-wind tour of the big ones, including climate change, in this chapter. Because of the scale at which our species has changed the earth, many consider the current time period as distinctive, perhaps so different from the past that it deserves its own geologic epoch – the **Anthropocene**. This term is a young one, first proposed in 2000, but like other terminology we've discussed here it has caught on for a reason. It encapsulates the sheer scale at which the earth has been altered by human activities. Interestingly, to technically categorize a new geologic epoch, earth scientists will insist that certain highly specific scientific criteria are met and in 2024 two dozen experts voted that we are officially *not* in a new geologic epoch[1] – hold your horses everyone, we are still in the Holocene (Zhong, 2024). Okay, fine, but culturally we are in the Anthropocene. This term caught on not because people were concerned about geologic strata or scientific criteria. They were slack-jawed and somewhat horrified by the sheer scale of the change that one little species, good old *Homo sapiens*, has made to its home planet. Over just a few human generations, in the blink of an eye on a planetary time scale, everything is different. I agree with Yadvinder Malhi (2017) who suggests we should judge this term not so much by its geologic accuracy but by its utility. Like other newly minted words such as biodiversity and ecosystem services, Anthropocene is a very useful concept. This word caught on for a reason. As Malhi writes, "it presents a new and potent framing for thinking about the unique challenges facing a burgeoning humanity on a finite planet" (Malhi, 2017, p. 99).

Nine Planetary Boundaries: Biodiversity Loss, Climate Change, Nutrient Pollution, Novel Entities, and More

Just how much have we changed the earth and what risks are involved with these global changes? We do after all need an ecologically functioning planet. Just like other animals, we are adapted to live in this place at this time. We are dependent on nature and part of it, and we are dependent on nature staying a certain way. We would be hard pressed, for example, to live on young earth with just the prokaryotes to hang out with and hardly any oxygen in the air. We need plants to eat and to feed to animals, we need the air we breathe and the water we drink to be clean, we need healthy soils teeming with life. How do we get a handle on how the pressures we are placing on the planet are influencing the planet's life support system? A group of researchers has done work which can help us wrap our heads around this exact question and we'll use this work to guide us on our quick tour of global ecological change. Katherine Richardson and others (2023) have collaborated to address global change from the framework of what they call **planetary boundaries**. At present, there are nine global environmental changes that this work monitors to gauge where we are at in terms of pushing the planet past a safe zone and increasing the risk of "earth system destabilization." Take a deep breath and reread that phrase a couple of times, letting the words sink in. Whenever I talk about this research, I tend to try to work in some dark humor, "well that sounds bleeping awful, can we please avoid that?!" For each global change, these researchers assign a risk level based on current human activities: they determine a planetary boundary condition, rate for a process that what would keep conditions in a Holocene-like state, and then, if we are below that boundary we are in the safe zone (in Figure 7.1 this is the inner patterned circle which is labeled "safe operating space"). Good job, humans! Once the boundary has been passed, however, the risk of destabilization increases and that risk increases as our activities push our planet farther past the safe planetary boundary.

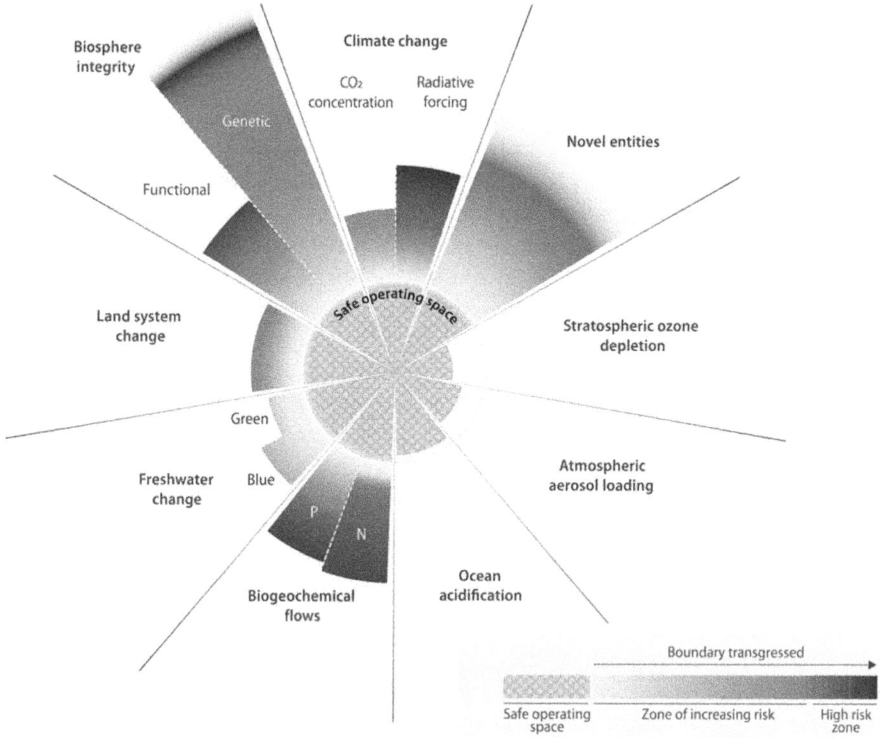

Figure 7.1 Current state of the planet for nine planetary boundaries
Note: The inner patterned circle shows the "safe operating space." As of 2023, six of the nine boundary conditions are currently outside of this circle including four which are in the danger zone: climate, nutrient pollution (labeled biogeochemical flows), novel entities (such as plastics), and biosphere integrity (e.g., biodiversity loss).
Source: Reprinted with permission from Richardson and others (2023).

With increasing risk, the wedges extend farther and farther from that inner circle. For example, loss of genetic diversity is put in the highest risk zone (Figure 7.1). To translate this from calm, level-headed science language to the everyday vernacular we might think of the high-risk zones in Figure 7.1 as shouting out "Stop! Now! What the heck are you doing you silly hubristic species?! This is your home!"[2] Richardson's research places our planet outside of the safe zone for six of the nine global changes including, of course, climate change. This bears repeating – the climate crisis is just one of six global-level environmental changes that are forcing conditions outside Holocene-like states. The scientists who do this work emphasize the importance of viewing the planet as an interacting system rather than thinking of global problems individually. So, what are these six big global changes? The four boundaries that show the highest level of risk are 1) climate along with 2) nutrient pollution (reactive nitrogen and phosphorous), 3) novel entities (plastics, forever chemicals, and more), and 4) biosphere integrity (deforestation, loss of genetic diversity). Why don't we hear more about global changes that aren't climate change? Jan Konietzko has called this **carbon tunnel vision** – our tendency to focus solely on atmospheric carbon and the climate crisis – despite myriad other global environmental challenges (Konietzko, 2022). Global change of course includes climate change but is not limited to it. Humans have fundamentally changed our home planet by moving species around (invasive species), manufacturing synthetic chemicals and other novel entities (plastic and other synthetic chemicals), altering biogeochemical cycles (increasing reactive nitrogen and phosphorous in water systems), changing land use (deforestation), changing the atmosphere (which influences a lot more than climate), and, finally, yes, changing global climate. Phew, welcome to the Anthropocene indeed! Buckle up for the ride, buttercup. If we want to keep our planet in the zone where *all past civilizations have occurred* (Holocene attributes), we have our work cut out for us! Let's look at each of these global processes and their planetary boundaries individually so we can familiarize ourselves with the grand challenge facing us. I'll do this in the order of bad news to good news starting with the four global changes that currently place the planet at high risk of destabilization (deep trouble, Figure 7.1) and ending with global changes which are closer to, or in some cases within, the Holocene planetary safe zone. We have, believe it or not, sometimes gotten things right and that deserves some celebration and a call of "nice job, humans." Specifically, we passed the Montreal Protocol and controlled CFCs! We will see that stratospheric ozone depletion is one risk we have dealt with effectively by passing an international protocol that controlled the chemicals that degrade the ozone layer. Fantastic, do it again!

Biosphere Integrity – Biodiversity Loss

In Chapter 6, we looked at biodiversity loss and saw that the current rate of extinction greatly outpaces the background rate found in the fossil record. In fact, this rate is actually faster than the extinction rates that lead to the five mass extinction events in earth's history. Losing species at this rate is very risky. Aldo Leopold famously wrote, "To keep every cog and wheel is the first precaution of intelligent tinkering." In terms of the planetary boundary, we are at high risk (what I like to call – please stop now) for biological integrity, with both the loss of genetic diversity and the percent of the biosphere appropriated by human activities putting the planet's stability at risk (Figure 7.1). There are several factors, what researchers call drivers, that are leading to this extreme rate of biodiversity loss. People are changing places that used to have grasslands and forests into agricultural fields and plantations, that is, taking the **primary production** that used to fuel the wild food chains and using it to fuel our material needs and desires. Biodiversity is lost when this occurs. Habitat is destroyed or degraded. Crossing other planetary boundaries also fuels biodiversity loss – pollution and climate change degrade habitats. We also move species around the planet and some invade areas, displacing the native species that lived there historically. Species are also hunted and caught as a byproduct of our fisheries (bycatch). Bycatch refers to mortality in non-target species from fishery activities. Marine vertebrates as a group, including dolphins and sea turtles, get caught in nets that companies use to catch fish. Bycatch is a consequence of global-scale fisheries and a major cause of marine mammal and reptile declines. In part this is due to the fact that the animals that get caught are the larger, older individuals in the populations. Older individuals are often critically important for keeping population numbers stable. Using sea turtles as an example, adult turtles lay a lot of eggs, so adult survival is key to maintaining population numbers. We know, from demographic studies, that even if 100% of the eggs survived, populations of loggerhead sea turtles would still decline unless survival is increased for older turtles (Crouse et al., 1987).

Richardson's planetary boundary work uses extinction rates to gauge our impact on the biosphere along with a variable called NPP (net primary production). Net primary production is the amount of plant biomass the earth generates over a period of time. A large percentage of the earth's NPP is now harnessed by people (e.g. croplands). As humans use more of the earth's surface for our own activities, habitat for other species declines. Long before a species goes extinct it experiences population declines. One by one local populations disappear leading to an overall decline in numbers. Associated with these population declines is a decline in genetic diversity. In addition to ecosystems getting replaced outright by human development, existing ecosystems also get degraded in quality with invasive species and pollutants making them less suitable habitat over time. Not surprisingly these changes affect the ecosystem services these systems supply. To prevent extinction, species need to have large populations. Specifically, we want to avoid what biologist's call the **extinction vortex**. The extinction vortex refers to the fact that when species reach small numbers several factors interact to drive their population further down in size like a whirlpool sucking a species to eventual extinction. One of these factors is **inbreeding depression** which occurs when related individuals produce offspring leaving offspring with two copies of deleterious alleles. This is how you get rhinoceroses with just one ear, Florida panthers with kinked tails, and declining reproduction in Prairie chickens. You might wonder why such "bad genes" are still floating around in populations. If these genes are harmful, why do species still have them? Part of the answer is that these deleterious versions of genes tend to be recessive in nature. It is therefore easy for an individual to carry a copy with no associated problems. The dominant allele is fine and codes for a gene product that works. Unfortunately, at small population sizes individuals that remain start to be close relatives. Close relatives are more likely to share harmful, recessive alleles and inbreeding depression then occurs in their offspring that are saddled with two bad versions of a gene. Ideally, this extinction vortex is avoided altogether by keeping population numbers, and associated genetic diversity, high. So, how healthy are the population numbers of various species across the globe? It depends on what species we look at. The Living Planet Index[3] database offers a great resource in this respect with detailed information on population changes of species for which we have good data. The news is mixed, with many populations declining in number but with some common species showing stable or increasing population sizes.

Even more troubling perhaps than genetic or species diversity declines, is the decline in whole ecosystem types (Watson et al., 2016). Globally, a little more than half of all land is human-dominated. But land conversion is not uniform across different **biomes**. Human preferences mean some biomes show low rates of habitat loss, for example, tundra and boreal forest. These biomes are less hospitable to human life and its associated activities which has meant lower rates of habitat conversion. In contrast, we have biomes from milder climates that are at much greater risk. Grasslands are particularly susceptible to decline because people convert grasslands to agriculture. Dry tropical forests are another ecosystem experiencing dramatic loss. Ideally, all biomes would have some base level of protection both to stop biodiversity loss and, also, to confer local landscapes with associated ecosystem services.

Invasive species are also playing a strong role in biodiversity decline. Humans have gotten very good at moving material around the planet. This is leading to homogenization of the earth's flora and fauna; ecosystems that used to look quite divergent, because of the species that lived in them, are starting to look homogeneous. People are sometimes shocked to learn that certain species we are quite fond of are **nonnative** to a given region. Throughout history people have often very deliberately taken species they like, such as honeybees, with them on their travels to new places, or introduced species they like (yum, trout) into areas those species weren't historically found. Often, however, species make it somewhere new on their own – in the ballast of ship water, burrowed in a piece of lumber, or some other way. Most of these stowaways find themselves in an unhospitable habitat with few other individuals to mate with; they die without reproducing and nothing changes. Very rarely, however, something quite different happens. The new habitat is great. In fact, it's better than where they came from because this new place doesn't have the predators and diseases. Reproduction and survival take off, and populations of a nonnative newcomer explode. These are the species we call invasive and they tend to make headline news because they become so numerous that they start to cause more and more headaches both for folks tasked with managing ecosystems and for folks in other industries. Zebra and quagga mussels, Emerald Ash Borer, Amur's Honeysuckle, Purple loosestrife, Lionfish, Asian Carp, Burmese python, Oriental bittersweet … to name but a few North American invasive species. Such species can have devastating impacts on the biodiversity of the ecosystems they invade. Let's take just one species as an example. Burmese pythons are a nonnative, invasive in Florida – establishing themselves in ecosystems that did not historically have any predatory snakes of this size. The novelty of this new apex predator has had major impacts on Florida's mammal community (Dorcas et al., 2012). Michael Dorcas and others (2012) showed that python invasion resulted in dramatic declines in the abundance of raccoons, opossums, rabbits, and bobcats. Given what we know about **interaction webs** (Chapter 5), it isn't difficult to predict that changes in these mammal populations are affecting the rest of the Everglades ecosystem. Invasive species like these pythons frequently cause dramatic shifts and declines in biodiversity. Biodiversity loss is just one, though, of four planetary boundary conditions that are in the danger zone. There are three other global changes that also rank as high risk so let's move on from predators to pollution, specifically, nutrient pollution.

Eutrophication – Nitrogen and Phosphorous Pollution

By now most people are well aware of how humanity has altered the carbon cycle, and we will get to the climate crisis and ocean acidification soon enough, but what about other biogeochemical cycles? Have our activities affected them and do those alterations have any associated risks? Brace yourself for some bad news. We have significantly altered both the nitrogen (N) and phosphorus (P) cycles with particularly dramatic impacts to aquatic ecosystems both freshwater and marine. Put simply, our changes to these cycles put water-based ecosystems at risk. Just like with biosphere integrity (biodiversity loss), nutrient flows are a global-scale change pushing our planet into the high-risk zone (Richardson et al., 2023; Figure 7.1). If you have ever lived by a lake undergoing an algal bloom you may have experienced the negative effects of nutrient pollution first hand. You may even have had your water supply affected. Not only can an algal bloom shift a recreationally pleasant lake to a slimy, mucky place, in some instances, blooms endanger local water supplies because certain types of algae secrete toxins. Let's look at both of these nutrient cycles and the changes we've caused in them. We'll see that for both cycles (N and P) our impact is one of ecosystem fertilization – plants and algae grow better when you add fertilizers – and nitrogen and phosphorous are just that.

Nitrogen is the most abundant gas in the atmosphere. Take a deep breath – 78% of what you just had in your lungs is nitrogen in its most inert form (gaseous nitrogen – N_2). There is *so much* nitrogen in the air, *but* it is not very useful in this form to the vast majority of living things. The two atoms in N_2, gaseous nitrogen, are held together *incredibly* tightly by a strong triple bond. This means it takes a lot of energy to separate those two nitrogen atoms from one another. Humans, however, are clever and good at breaking things, and we figured out how to do just that – break that triple bond and add some hydrogens on to each individual N atom. Agricultural fertilizer (liquid ammonia) is created by putting nitrogen and hydrogen gases together under high heat and pressure. This process, the Haber-Bosch process, was refined by chemists in the early half of the 20th century. Now, in my home state of Illinois, every autumn after the corn and soybeans are harvested, large white metal tanks are pulled out onto commercial farms, and liquid ammonia is injected into the fields. This fertilizer helps increase crop yields the following growing season. Ammonia is one nitrogen atom with three attached hydrogens and it quickly interacts with other molecules in the soil to make a variety of nitrogen forms that we'll refer to collectively as **reactive nitrogen**. Burning fossil fuels also creates reactive nitrogen molecules. Reactive forms of nitrogen have single bonds with either hydrogens or oxygens and are much more, you guessed it … reactive than gaseous nitrogen. Reactive nitrogen is good news for plants because it means they can easily suck nitrogen up into their hungry roots and, because these forms are chemically reactive, they can use this nitrogen to build proteins and other important pretty plant parts.

The phosphorous cycle is different than nitrogen cycle in several ways. One major difference is that, unlike nitrogen, we don't find phosphorous in the air except as dust. Phosphorous doesn't have a gaseous form. The way we use phosphorous, however, is similar to our use of reactive nitrogen. Like nitrogen, phosphorous is a great fertilizer. Because of this, we mine it and spread it on fields to make crops grow. Phosphorous is also found in household products including laundry and dishwasher detergents. In fact, the amount of phosphorous in detergents is regulated by state law in many US states. This is done to help decrease the amount of phosphorous in waste water. Nitrogen and phosphorous are also both found in manure. What all this means is that as we have more people washing their clothes; growing food; driving cars; raising livestock; the levels of reactive nitrogen and phosphorous in nearby air, land, and water increase. We've altered these nutrient cycles in no small part because these compounds are terrific fertilizers and making plants grow big is beneficial from farms to flower gardens. But when these fertilizers reach a local stream, lake, or estuary, they don't stop doing what fertilizers do. They continue to make photosynthetic species grow, except in waterways this includes making phytoplankton grow, and the result can be an algal bloom.

Eutrophication is what happens to a body of water when it has excess levels of nutrients (N and P). Depending on the type of water body, say a freshwater lake versus an ocean bay, phosphorous or nitrogen may be the main nutrient leading to algal growth. However, the overall changes caused in an ecosystem undergoing eutrophication are the same regardless of which particular nutrient causes the shift towards algal dominance. Let's walk through what happens when a lake becomes eutrophied with the understanding that we could just as easily have picked an ocean anywhere a large river meets the sea. Runoff with high nutrient levels flows into the lake and the lake's algae experience a fertilization effect and begin to grow with reckless abandon and over time – if the nutrient flow keeps up – a thick layer of green muck will grow on the lake's surface. Picture in your mind a slimy lake with a thick mat of pale green sludge on top. Along with this layer of algal growth, there is an associated increase in dead algae. This means there is a lot of dead stuff (technically called dead organic matter (DOM)) for decomposers to break down. The process of decomposition uses a lot of oxygen so as there is more dead material, the oxygen levels in the lake start to decline. In extreme cases of eutrophication, oxygen is being used up so quickly you can't even detect it in the water. This condition is of course lethal for many of the species that live in the lake. For example, all the fish that get their required oxygen from the water using their gills. Unfortunately, this starts a run-away effect. As the fish and other animals in the lake start to die from the low oxygen levels, there is further build-up of dead material which creates an even greater demand for oxygen by the decomposers. Eutrophication can cause what ecologists call a **regime shift** – a shift from a clear lake with rooted plants in the lake bottom to an algae-dominated eutrophied lake. Making matters even worse, some algal blooms, labeled harmful algal blooms, release toxins in the water making the water unsafe for use. In addition, as water warms it is more susceptible to algal blooms so climate change increases the likelihood of blooms occurring. This brings us to our next global change. Hopefully, you are starting to see how these changes are interconnected systemically; more farm fields means less forest and biodiversity, and more fertilizers going into warmer lakes which increases algal blooms. Lakes and oceans are heating up as we heat up the planet. Which brings us to the big C: climate.

Climate Change

People have altered the earth's atmosphere. This alteration includes increases in the big three **greenhouse gases**: carbon dioxide, nitrous oxide, and methane. All of these molecules absorb heat as it comes off the planet. It makes these molecules wiggle and thus reradiate heat. It is actually good that earth has low levels of greenhouse gases in our atmosphere, without them it would be an incredibly cold planet and we wouldn't probably be here. The background, Holocene, level of carbon dioxide was 270 parts per million (ppm), currently, in 2026, that value is over 400ppm, and we are already experiencing climate shocks including devastating hurricanes and wildfires. Climate influences all the ecosystems on earth. When ecologists characterize **biomes**, we do so based on an area's regional climate and dominant vegetation. The dominant types of plants that are found in any given region varies depending on the region's annual pattern of rainfall and temperature. Perhaps the easiest way to think about climate is to climb a mountain in our minds. Because of their abrupt elevational changes, mountains are unique in having multiple climates very close to one another. As one treks up the side of the mountain you might walk from a scrubby ecosystem with grass and a few bushes up through different forest types and, if your mountain is tall enough, perhaps even up to a tree-less area where the climate is too cold to sustain trees. Because mountains have this unique property – many climates close together – they are good locations for conserving biodiversity as the global climate shifts. As the earth warms and mountain climatic zones shift upwards, plants and animals can easily track these shifts due to the short distances involved. Of course, if plants and animals are adapted to the climate only found at the mountain's peak they are at risk as this climate disappears and they have nowhere to go.

How will earth's systems respond to changing climate? How will we respond? To put climate change in context let's start with some baseline statistics. The average temperature of the planet is 15°C (59F). This is a *global* average across the entire earth. The planet has gone through periods where it was both hotter and cooler than it is at present. This is due to long time scale Milankovitch cycles. These cycles happen *incredibly* slowly compared to current anthropogenic climate change but they serve to orient us in thinking about what a hotter and cooler planet look like. How much colder was the earth 10,000+ years ago when glaciers roamed North America? 10°C on average so 5°C colder. I find it useful to keep this average in mind when international groups discuss global goals under current climate change, such as keeping the earth no more than 1.5°C degrees hotter than the Holocene climate. We have already seen a 1.2°C increase in average global temperature. These numbers can seem small but it is because we live our lives, and therefore think, in local temperatures not global averages. To use an analogy to explain the difference in scale, think about the food you would have to personally eat to gain 5 pounds. Now think of how much food everyone on the planet would have to eat to raise the average weight of 8 billion people 5 pounds. It is the same when we think about increasing the average global temperature by a degree, a *lot* of increase in radiative forcing is needed to move the needle. This also means that an earth that is just 1°C degree warmer *on average* is a fundamentally different system. Ecosystems are changing on our warmer planet with earlier spring events, animals and plants moving their ranges towards the poles, and shifts in local ecosystem types and communities towards heat-tolerant species. So many scientists have covered the climate crisis so well that I will just refer you to the many reports that discuss what we are already seeing and what we should expect under different carbon control scenarios. For the US, the National Climate Assessment Report is wonderful in providing specific information on changes we have seen and expect throughout my country. Unfortunately, this national report which is required by congress has faced increasing political pressure with the Trump Administration firing the authors of this report in April 2025 (Volcovici, 2025). Globally, we of course have all the incredible work of the Intergovernmental Panel on Climate Change (IPCC).

The thing I find most terrifying about climate change is that all of human civilization – the past 10,000 years – has occurred during an incredibly stable climatic period of earth's history. The wars, the genocides, all of history from the Greeks and the Egyptian pyramids until present day happened in the context of a stable climate. I worry more about what climate shocks will do to our social systems than what they'll do to our ecosystems. Of course, the social and ecological systems are related as loss of ecosystem services and regime shifts puts pressure on social systems. Nature, however, is resilient over geologic time. Biodiversity has taken huge hits before and slowly but surely clawed its way back. I'm so fond of (most) people though and I truly want to see us get this right.[4] This includes getting carbon out of the atmosphere and back into the soil and plants. Part of how we do that is by increasing the land we dedicate to natural **carbon sinks** – increasing rather than decreasing global forests and grasslands. Natural ecosystems play a significant role in affecting local, regional, and global climate. The Amazon forest is perhaps the best known in this regard (Box 7.1).

Box 7.1 Amazon forest tipping points: Water stress and risk of regime shift

Headlines about the Amazon forest can be terrifying: fires, drought, deforestation, and the auspicious phrase "tipping point." News and policy often hint of a point of no return where the Amazon has been so altered by human activities, including climate change, that it will shift abruptly to a different **biome** altogether – perhaps savannah with fewer trees – such a shift would have devastating impacts both regionally and globally. What is the science behind these headlines? Let's start with what we know about the Amazon forest. It covers a large proportion of South America including nine countries – the majority of the Amazon, 60%, is in Brazil. This region of South America has remained forested for 65 million years despite large shifts in global climate during that period. We also know that this incredible tropical forest contains at least 10% of the world's terrestrial biodiversity, stores a massive amount of carbon, and has a cooling effect because of all the water moving through the trees. The Amazon creates rainfall regionally and is therefore important for water availability across South America. Brazil specifically supplies a large percentage of the rainfall to western Amazonian countries. Bottomline – we need the Amazon to remain forested and this forest faces several threats that combined often lead to water stress. These threats are also well known: climate change, deforestation, prolonged drought, wildfires, and invasive grasses. As roads cut into the forest and as logging occurs, remaining smaller forests are more susceptible to drying, burning, and invasion. These threats are greatest along the southeastern edge of the biome – ominously labeled the "Arc of Deforestation." This southeastern region has already shifted away from absorbing carbon dioxide and is now a source of greenhouse gases. The current threats are real and daunting but are we at a point of no return or can the Amazon be saved?

Paulo Brando and others (2025) set out to answer just this question by reviewing scientific papers on Amazon forest degradation. The conclusion they reached in their review is simultaneously hopeful and frightening. The hopeful piece is that there isn't strong evidence of what they refer to as a "single, basin-wide tipping point" (Brando et al., 2025, p. 120) or what other authors call "large-scale collapse of the Amazon forest" (Flores et al., 2024, p. 562). Brando and others (2025) call this tipping-point concept a domino hypothesis – where once one domino falls, the fate of the others is already determined. Deforestation leads to drier forest, which leads to burning, which leads to more invasion, and so forth. A cascading set of events that pull the forest away from its present state. In contrast to this domino hypothesis, Brando proposes a hammer hypothesis where the forest continues to degrade, and its resilience decline, because people keep banging on it with various hammers (drier climate, logging, invasive species, fire). Support for this second hypothesis is strong with forest areas losing resilience as threats increase. To build back resilience in this critically important ecosystem, several things need to happen. At a global scale, the climate crisis needs to be addressed aggressively. Within the biome itself, deforestation needs to stop and afforestation increase. Strong environmental governance is also needed to manage wildfire and illegal logging. It is also critically important that protected areas and Indigenous territories be respected and, whenever possible, increased in size. It is these regions that currently have the highest resilience with little logging or fire. The task is different in different regions with afforestation being critical in the impacted southeast. As Brando and others write (2025, p. 122), "A critical window of opportunity remains open for safeguarding Amazon's resilience: Halting deforestation, managing fire, securing Indigenous land rights, and scaling up ecological restoration are viable and proven strategies." Hammers need to be put down and shovels lifted. When in doubt, plant a tree.

Novel Entities – Synthetic Chemicals, Pervasive Plastics, and More

Carbon dioxide, reactive nitrogen, phosphorous all of these can be considered pollutants at least in certain situations where they produce undesirable outcomes. In addition to altering our biogeochemical cycles, we have created substances that are completely novel. Our understanding of chemistry has allowed us to create insecticides, surfactants, plastics, all manner of brand-spanking-new to the planet substances. Determining the impact these substances have is difficult and, unfortunately, we often learn by observing the harm first hand. At best, safety testing prior to release looks at acute toxicity so other impacts are typically learned once products are used. For example, it took plummeting bald eagle populations to learn that dichlorodiphenyltrichloroethane (DDT) was detrimental to birds of prey and even further research to find out why: that DDT is an estrogen-mimic (endocrine disrupting chemical) that led to softened eagle egg shells. We learned that chlorofluorocarbons (CFCs) destroyed stratospheric ozone *after* the ozone hole appeared. There is always an inherent risk in introducing what Richardson and others (2023) call **novel entities** to the planetary system. One of the most prevalent such entities now is synthetic plastics.

Have you checked the label recently on that soft, fuzzy blanket? That cute sweater you picked up for a great price? If you're a bargain shopper, the label might surprise you: 100% polyester, otherwise known as plastic.[5] Cheap for us, increasingly expensive for the planet. Fast fashion and home décor are just two industries contributing to global plastic pollution. As an ecologist, I worry the most about any novel materials, including plastics, that don't decompose. Because it is cheap to manufacture and use for packaging – plastic is quite literally *everywhere*, including in many single-use items. Use it once and toss it, but then where does it go? The problem with plastic is it is not a natural material. Unlike wood or other plant fibers, it can't be decomposed by bacteria or fungi.[6] This means plastics build up on the planet over time. It can weather, which means breaking down into smaller and smaller pieces, creating in the process **microplastics**. But just because we stop seeing this plastic, doesn't mean it isn't still there. Weathering is not decomposition. Unlike decomposition, this process does not result in nutrients that feed new life. It doesn't change the chemical nature of the materials. It just makes them invisible to the naked eye. Because of the prevalence of plastic in our lives, microplastics are now found just about everywhere on earth including inside our food, drink, and bodies. Impacts of plastics on human health are still unknown but we shouldn't conflate uncertainty with safety. Huge areas of microplastic soup swirl in our earth's oceans in areas where there is confluence among the ocean's currents – the so-called plastic patches. This increasing concentration of plastic can be deadly to the ocean's magnificent creatures which can mistake little pieces of plastic as food (plankton) or become entangled in plastic netting. It's important as we think about sustainable futures to shift our consumption patterns to materials that nature can handle efficiently. When in doubt, pick the material that decomposes. Ecosystems are incredible – they can take a dead coyote and turn it into soil, air, nutrients and new life. Pushing up daisies from the detritus. Don't even get me started on what a magnificent thing a dead whale at the bottom of the sea is. Okay, I can't help myself, quick aside. Whale carcasses have been found to be of fundamental importance for marine biodiversity. Referred to as "whale falls," a nutrient-rich whale carcass supports a teeming ecosystem on the sea floor (Kwong, 2019). Now back to decomposition in general. Decomposition writ large is an incredible service (nature's recycling), but we can't harness it unless we make items out of materials that nature recognizes. Specifically, materials that species have evolved to snack on. Novel entities need to be reined in and better tested before used. With these, as with the other changes we've already discussed, we are in the high-risk zone. There will no doubt be cases where the benefits of continuing to produce certain products outweigh the risks but assuming synthetic materials are safe until proven hazardous is naïve given what we know about how the earth works. Much better safety testing and life cycle analysis is needed for all novel entities. In discussing novel entities, we have now made our way through all four of the four planetary boundaries where our activities put our home planet at deep risk (Figure 7.1), time to look at the five boundaries where the news is a tad bit better.

Box 7.2 Bioplastics: Are they a solution to plastic pollution?

Aquatic ecosystems are particularly at risk from plastic pollution. The base of the food chain in rivers, lakes, and the ocean consists of tiny, floating creatures: first the group with the gift of photosynthesis, phytoplankton, and then the little animals that munch on them, zooplankton. Sadly, a small floating piece of plastic often has a similar size and shape as these plankton which explains the fact that an increasing percentage of the diet of aquatic animals is now just that: plastic. These tiny plastic bits, **microplastics**, also tend to bind toxic chemicals and they accumulate in the food chain affecting the health and survival of top predators such as fish or marine mammals. I'm sure that if we were fish, we would have developed better plastic policies by now but even us land lovers are increasingly exposed to this prevalent pollutant. Industries that rely on plastics are well aware of the problem and are working to develop alternative materials with similar properties. This is where **bioplastics** come in. These are a wide range of materials that fall into three big groups: 1) derived from biological material and biodegradable, 2) derived from fossil fuels and biodegradable, and 3) derived from biological material but not biodegradable (Moshood et al., 2022). All of these are at least somewhat better for the environment than synthetic plastics which are derived from fossil fuels and aren't biodegradable, but none are a silver bullet. This might come as a surprise. At first glance, group one sounds perfect: industry mimicking nature. One example from this group would be starch, derived from corn or potatoes, as a starting material used to create a pliable substance that can serve a given function and then when it enters the waste stream, decomposition breaks it down. Theoretically, this is a perfect material and it would be if we were back in the 1950s developing plastics for the first time. But it's not the 1950s, so when new materials are created we need to think about how they will fit into current global systems which include plastic policies, consumer behavior, and waste disposal.

Grab a plastic bottle and look at the bottom of it. You'll probably find a recycle symbol with a number inside. Different types of plastics need to be recycled in different ways. However, to get recycled someone has to put the item in a recycling bin. In turns out only 7% of plastics are recycled (Thacharodi et al., 2024). Most plastics end of in the general waste stream destined for a landfill. What happens when we start to increase bioplastics in this system? We can anticipate that a majority, at least at first, would end up in landfills. This is where the interconnected nature of global environmental problems raises its ugly head. Deep down in landfills there is little oxygen and in such environments the bacteria that break down organic material produce methane as a byproduct. Methane is unfortunately a potent greenhouse gas with 23 times the heat absorption power of carbon dioxide. Ramping up bioplastic production without changing land fill design, consumer behavior, and policies could end up increasing methane release from the world's landfills exacerbating climate change. This means shifting to bioplastics comes with a set of challenges including increasing the percentage of landfills that can capture methane, good labeling and changes in consumer behavior so bioplastics are processed appropriately post-consumer, and creating the policies that go along with forcing these shifts.

Land Use and Freshwater Change

Think of a location on the planet's surface and envision yourself there. Keep it simple if you like and just look around the place you find yourself in as you read this sentence. Now imagine the same place a thousand years ago. For me, in Illinois, ironically called the Prairie State, it is hard for me to wrap my mind around the sweeping prairie (sea of grass) that could once be seen in every direction. Now I stand on a paved street, my house with a pollinator garden out front is nested in a small city surrounded not by prairie but by fields and fields of corn and soybeans. My students sometimes call this area of Illinois the "corn desert." Ecologists call this change in what types of ecosystems are at a given location – **land use change**. This same geographic location which historically was tallgrass prairie is now agriculture, primarily. As people dominate landscapes, other ecosystems decline and urban and agroecosystems expand. Forests (temperate to tropical), grasslands, wild running streams and rivers, ocean bays and estuaries, rocky intertidal pools. In addition to being home and habitat to the earth's biodiversity, these ecosystems benefit us and the planet in a variety of ways. We need to keep a good chunk of systems in their natural state, particularly our global forests, because of the role natural ecosystems play in regulating the earth's systems. Global deforestation rates have declined but have not gone to zero. This means we are still losing forest globally. Ideally, we would not just stop global loss but start increasing forests and grasslands due to the benefits they provide and the species they protect. Forests are **carbon sinks** – absorbing atmospheric carbon and thus helping to control climate. Increasing wildfires due to water stress and disease are making forest preservation ever more challenging. Climate change has led to warmer winters; bark beetles can now survive the winter and attack trees, increasing dead trees on the landscape and fueling future fires. As forests and grassland dynamics shift, keeping more freshwater in the soils, groundwater, and streams becomes increasingly important. Freshwater demand can be extremely challenging in arid regions where people's desire for water may be greater than the supply. In some locations, communities will have to make some tough decisions about the type of ecosystems they work to maintain. This is because more and more ecosystems are becoming what ecologists call **novel ecosystems**. These don't look like the ecosystems we've had in a given area in the past and their dynamics reinforce their novel characteristics (Hobbs, Higgs, and Hall, 2013). Richardson and others (2023) put changing land use outside of safe operating conditions but not as extreme a risk as the changes I covered earlier (Figure 7.1). They also, quite appropriately, point out the interrelationships among global changes. Part of solving global disruptions that are high risk (climate change, eutrophication) will involve maintaining natural systems and increasing forests, grasslands, and flowing rivers. These changes help decrease biodiversity loss and help combat climate change. Having natural ecosystems along rivers and streams keeps nitrogen and phosphorous from reaching aquatic ecosystems thus decreasing eutrophication. Part of thinking like an ecologist is realizing it truly is all interconnected. This isn't just a warm and fuzzy notion – it is what the science of global change has taught us.

When we think of water use, we may not readily see the pressure human activities place on the world's water resources. Our lifestyles use much more water than we realize because of the high amount of water use involved in energy production and manufacturing products (**hidden water**). Hidden water refers to this less visible water use. A consumer economy is one with high water use. Some crops, like cotton, are water thirsty. One cotton T-shirt takes 650 gallons of water to make. Water footprint calculators allow individuals to get a better understanding of hidden water because people can estimate their actual water use based on your lifestyle (Ecorise, 2022). In the next chapter we'll discuss sustainability and examples like this make it clear that a big piece of the answer is to make stuff well (durable), use it as long as possible, mend it, fix it, swap stuff when you're bored with what you have, and when it can't function in one way, upcycle the material to create something new. Less is more. More creativity and less consumption. In addition to using goods differently, items need to be made of natural materials so the item can decompose when it eventually ends up as waste. The planetary boundary work puts freshwater change which includes changes in streams, soils, and groundwater at intermediate risk (Richardson et al., 2023; Figure 7.1). Whether we are thinking of land or water, the changes we have made are outside of safe conditions which means restoring lands to natural states and keeping water in these natural systems is vital.

Altered Atmosphere: Ocean Acidification and More

We tend to associate the changes in earth's atmosphere with climate change. However, changes in atmospheric chemistry and composition affect our oceans and lands in many other ways. The world's oceans serve as a huge carbon sink; they absorb tons of carbon dioxide from the atmosphere and dissolve it into their waters. This happens because carbon dioxide reacts with water and creates a weak acid: carbonic acid. The good news is this carbon isn't in the atmosphere acting as a greenhouse gas and heating up the planet. The bad news is this acid is decreasing the pH of the ocean with consequences for marine life. Remember, the pH scale measures acidity levels and goes from 1 (very acidic) thorough neutral (7) to extremely basic (14). Our oceans are naturally basic with a pH around 8.2. The myriad species that inhabit our oceans – corals, mollusks, plankton, sea stars, jellies, fish, whales – have evolved in and are therefore adapted to the past environment. Decreasing pH changes ocean chemistry and, in particular, can impact rates of calcification. Hard corals are known for their hard "skeletons" which are made out of calcium carbonate. Lower pH waters can decrease thickening of these coral skeletons and make them break more easily. Just like losing calcium in our bones leads to osteoporosis, so changing carbonate chemistry leads to weaker reefs. Coral reef preservation is a fabulous stretch goal for our generation. This is because preserving coral reefs entails reining in so many other global excesses. You can't protect the reefs just by legally protecting local reef systems. Multiple, global-scale stressors lead to **coral bleaching** and reef loss including climate change (warmer oceans, sea level rise), **ocean acidification**, and eutrophication (coral get shaded out by algal mats). Because many of these are global-scale threats, the changes we need to make to preserve the reefs are global as well as local in nature. Making these changes will save so much more than the reefs themselves. Let's do it, people!

Carbon dioxide increases are just one of many human-caused changes to the atmosphere. Another change is higher amounts of reactive nitrogen gases (NO_x). This increase means that rainwater tends to have a little bit of fertilizer in it because these gases dissolve into the rain. In some cases, this is good news, forests can grow faster and suck in more atmospheric carbon if they are given a nitrogen boost. This makes them better carbon sinks than they would be in the absence of reactive nitrogen (de Vries, Du, and Butterbach-Bahl, 2014). In other cases, however, the news is not so good as weedy, nitrogen-hungry plants take over ecosystems where rare plants once thrived. Plants in certain ecosystems have adapted to historically low nitrogen conditions. Now, however, nitrogen levels are higher and these plants are outcompeted by nitrogen-hungry weeds. Unfortunately, this means nitrogen-rich rain is contributing to plant biodiversity loss (Payne et al., 2017). Ocean acidification and loss of biodiversity caused by nitrogen deposition highlight, yet again, why it is folly to approach global environmental issues in a piecemeal fashion. Changing climate is just one symptom of a dramatically altered planet. We need to address global environmental challenges holistically rather than take on each one individually. They are interconnected, wicked problems so their solutions also need to be interconnected. The whole planet is sick. I fear if we don't take this approach, we will forever be playing a game of whack-a-mole, addressing one environmental issue only to have a detrimental impact pop-up elsewhere. At an ecology conference over a decade ago in the southwest, I sat next to a tall, lean Native American man who made a comment that has stuck with me: "Mother Earth has a fever." We know with human health, treating just the fever is silly and masks the problem. It is important to determine what caused the fever and treat the root cause if the patient is to make a full recovery. I hope by now in this chapter it has become clear that climate change is just one symptom of the larger, more complex, and wicked problem of global environmental change (Figure 7.1).

A Little Bit of Good News

This chapter was tough to write and perhaps will also be hard to read. The numerous types of global changes and their interactions can leave even the most optimistic person feeling overwhelmed and powerless to make a difference. What can one person possibly do? We'll see however that we can stack solutions to tackle multiple global stressors simultaneously with single interventions. Another ray of hope can be found in history and – as hard as it may be to believe – global politics. We have created solutions to global-level disruptions in the past and these solutions provide us with a roadmap for the future. Let's wrap up this chapter with some good news which, believe it or not, involves the hole in the ozone layer.

The earth has several layers to its atmosphere. You and I hang out down here in the layer called the **troposphere**. Ironically, ozone[7] is a bad thing down here where you and I live and breathe. Ozone is created in the troposphere by fossil fuel burning and leads to poor air quality and associated health effects. This is particularly problematic in big cities which may have many cars and poor mixing of urban air. Several kilometers up into the atmosphere, however, ozone is a truly wonderful, awesomely important molecule. Stratospheric ozone. Way, way up, 20 kilometers above the earth's surface, the protective ozone layer begins and continues for about 10 more kilometers. This thick layer of ozone is incredibly important to life on earth because it is situated between the sun and our planet and blocks ultraviolet (UV) radiation. Most of us are well aware of the effects of high UV exposure, we have all been schooled in the need to apply sunscreen. Sunburns and cataracts are two examples of the damage UV can cause. What exactly happened to the ozone layer, how did it get its infamous hole? Novel entities were the culprit. Chloroflurocarbons (CFCs) are a group of synthetic chemicals that were first developed in the 1930s to use as cooling gases in refrigerators. Over the next four decades more uses for CFCs were found and, in many ways, they were thought to be a wonderful, new synthetic chemical group. Low toxicity, very inert. Better living through chemistry! They were however new to the planet (an early example of novel entities) and they had an effect on the planet as a whole that would have been hard if not impossible to anticipate. CFCs can be seen as a case study of why novel entities as a group are inherently risky. Novel entities are new to the planet and its ecosystems so it is impossible to determine every potential effect they may produce in these complex systems before release. In the case of CFCs, these chemicals were inert enough to float, seemingly innocently and harmlessly, farther and farther away from the surface of the earth. When they got several kilometers up, to the stratosphere, they encountered the ozone layer and on its far side, high levels of UV from the sun. This high UV was able to break off the chlorine atoms from these chemicals and this is where the trouble began. Just one small chlorine atom in the stratosphere can wreak a disproportionate amount of havoc – degrading on average 100,000 ozone molecules. Slowly, atmospheric scientists saw the ozone hole appear and then it got larger and larger. Fortunately, scientists were also able to determine what was causing ozone in the stratosphere to decline and they raised the alarm. What happened next is truly inspiring, although the politics were ugly as they often are, in 1987 the world's countries passed the Montreal Protocol banning global production and use of CFCs. This protocol is still in effect today and is in fact used to also combat the climate crisis as CFCs also turned out to be potent greenhouse gases. Figures, just our luck. If we go back to Katherine Richardson's work (2023) and check the figure from the planetary boundaries group (Figure 7.1), we find we are in the safe zone for stratospheric ozone! We actually did it! Scientists, activists, politicians came together across the globe to keep the planet in the safe zone. CFCs are banned and regulated via a global treaty, and the ozone hole has stabilized. Awesome job *Homo sapiens*, now do the climate crisis, deforestation, biodiversity loss, and eutrophication!

Conclusion

Earth is our home and we need to keep our home planet stable so that not only us, but *all* species, can live and flourish. It really is that simple. However, doing this will take more than just addressing the climate crisis. We need to actively work to restore all of earth's life support systems – preserving biodiversity; increasing forest and grassland cover; decreasing atmospheric carbon dioxide (critical for both ocean acidification and climate change); decreasing the flow of reactive nitrogen and phosphorous into our water bodies; and being much more cautious in our use of synthetic materials. It is going to take a lot, the change needed currently seems unimaginable. But imagine it we must. We have so much to lose if we don't make the necessary changes and so much to gain by trying.

Notes

1 In 2024, members of the International Union of Geological Sciences voted that it is wrong to classify the current era as a new geologic epoch (Zhong, 2024).
2 Scientists as a group tend to be understated in their phrasing so when it is appropriate, I'll try to translate jargon into everyday language which I've been advised

shouldn't contain expletives. I'll admit this is somewhat challenging in the face of certain data.

3 The Living Planet Index is a statistic which has shown a geometric average of 73% decline leading to stark headlines. This average however is hard to interpret in isolation so, like others, I'd suggest that if you are really interested in biodiversity loss at this scale you dive into the actual database that shows which populations of which species are declining where (Ritchie, 2024). The summary statistic alone masks a lot of variation.

4 Dr. Ayana Johnson's book *What if we get it right? Visions of climate futures* is a must read for anyone grappling with envisioning a positive future given the array of environmental challenges we face (Johnson, 2024).

5 Although plastic can be defined broadly to include all materials that can be heated and molded, including natural materials, I am referring to synthetic plastics here. Plastics in clothing come by many names: polyester, acrylic, nylon, rayon, spandex, and more.

6 Microbes have evolved to be able to breakdown very complex biomolecules such as lignin in plants. There is evidence of slow rates of decomposition of certain plastics by some microbes in certain environmental conditions (Sharma and Neelam, 2023). Again, however, rates matter, slow rates of decomposition can't keep up with the high rate of plastic pollution. Synthetic plastics are also a very broad group. The high molecular weight of many plastics makes them very stable in the environment. Understanding what plastic attributes and environmental conditions aid natural breakdown is increasing as is discovery of which strains of microbes can accomplish this task (Sharma and Neelam, 2023). Although by no means a silver bullet, nature may play a role as we work to get plastics under control.

7 Ozone is a form of oxygen – O_3. It occurs in a natural protective layer in the stratosphere. It is also produced in the troposphere when a single oxygen atom, O, is produced by burning fossil fuel which then reacts with O_2 in the air. This plays a significant role in creating smog.

References

Balint, P. J. (2011). *Wicked environmental problems: managing uncertainty and conflict*. Island Press.

Brando, P. M., Barlow, J., Macedo, M. N., Silvério, D. V., Ferreira, J. N., Maracahipes, L. … and Uribe, M. (2025). Tipping points of Amazonian forests: Beyond myths and toward solutions. *Annual Review of Environment and Resources*, 50. doi:10.1146/annurev-environ-111522-112804.

Crouse, D. T., Crowder, L. B., and Caswell, H. (1987). A stage-based population model for loggerhead sea turtles and implications for conservation. *Ecology*, 68(5), 1412–1423. doi:10.2307/1939225.

de Vries, W., Du, E., and Butterbach-Bahl, K. (2014). Short and long-term impacts of nitrogen deposition on carbon sequestration by forest ecosystems. *Current Opinion in Environmental Sustainability*, 9, 90–104. doi:10.1016/j.cosust.2014.09.001.

Dorcas, M. E., Willson, J. D., Reed, R. N., Snow, R. W., Rochford, M. R., Miller, M. A. … and Hart, K. M. (2012). Severe mammal declines coincide with proliferation of invasive Burmese pythons in Everglades National Park. *Proceedings of the National Academy of Sciences*, 109(7), 2418–2422. doi:10.1073/pnas.1115226109.

Ecorise. (2022, Oct. 20). *The hidden water in everyday products*. Water footprint calculator. https://watercalculator.org/footprint/the-hidden-water-in-everyday-products/.

Flores, B. M., Montoya, E., Sakschewski, B., Nascimento, N., Staal, A., Betts, R. A. … and Hirota, M. (2024). Critical transitions in the Amazon forest system. *Nature*, 626 (7999), 555–564. doi:10.1038/s41586-023-06970-0.

Hobbs, R. J., Higgs, E. S., and Hall, C. (2013). *Novel ecosystems: Intervening in the new ecological world order*. John Wiley & Sons.

Johnson, A. E. (2024). *What if we get it right?: Visions of climate futures*. OneWorld.

Konietzko, J. (2022, Feb 8). Moving beyond carbon tunnel vision with a sustainable data strategy. https://www.cognizant.com/us/en/insights/insights-blog/moving-beyond-carbon-tunnel-vision-with-a-sustainability-data-strategy-codex7121.

Kwong, E. (2019, Nov 7). What happens after a whale dies? National Public Radio. https://www.npr.org/2019/09/13/760664122/what-happens-after-a-whale-dies.

Malhi, Y. (2017). The concept of the Anthropocene. *Annual Review of Environment and Resources*, 42(1), 77–104. doi:10.1146/annurev-environ-102016-060854.

Moshood, T. D., Nawanir, G., Mahmud, F., Mohamad, F., Ahmad, M. H., and AbdulGhani, A. (2022). Sustainability of biodegradable plastics: New problem or solution to solve the global plastic pollution? *Current Research in Green and Sustainable Chemistry*, 5, 100273. doi:10.1016/j.crgsc.2022.100273.

Payne, R. J., Dise, N. B., Field, C. D., Dore, A. J., Caporn, S. J., and Stevens, C. J. (2017). Nitrogen deposition and plant biodiversity: past, present, and future. *Frontiers in Ecology and the Environment*, 15(8), 431–436. doi:10.1002/fee.1528.

Richardson, K., Steffen, W., Lucht, W., Bendtsen, J., Cornell, S. E., Donges, J. F. … and Rockström, J. (2023). Earth beyond six of nine planetary boundaries. *Science advances*, 9(37), eadh2458. doi:10.1126/sciadv.adh2458.

Ritchie, H. (2024, October). Living planet index: What does it mean? Our World in Data. https://ourworldindata.org/living-planet-index-decline.

Rittel, H. W. and Webber, M. M. (1973). Dilemmas in a general theory of planning. *Policy Sciences*, 4(2), 155–169. doi:10.1007/BF01405730.

Sharma, H. and Neelam, D. K. (2023). Understanding challenges associated with plastic and bacterial approach toward plastic degradation. *Journal of Basic Microbiology*, 63(3–4),292–307. doi:10.1002/jobm.202200428.

Thacharodi, A., Meenatchi, R., Hassan, S., Hussain, N., Bhat, M. A., Arockiaraj, J. … and Pugazhendhi, A. (2024). Microplastics in the environment: a critical overview on its fate, toxicity, implications, management, and bioremediation strategies. *Journal of Environmental Management*, 349, 119433. doi:10.1016/j.jenvman.2023.119433.

Volcovici, V. (2025, April 29). U.S. dismisses all authors of national climate assessment: email says. *Reuters*. https://www.reuters.com/sustainability/cop/trump-administration-dismisses-all-authors-key-climate-report-email-says-2025-04-28/.

Watson, J. E., Jones, K. R., Fuller, R. A., Marco, M. D., Segan, D. B., Butchart, S. H. … and Venter, O. (2016). Persistent disparities between recent rates of habitat conversion and protection and implications for future global conservation targets. *Conservation Letters*, 9(6), 413–421. doi:10.1111/conl.12295.

Zhong, R. (2024, March 20). Geologists make it official: We're not in an 'Anthropocene' Epoch. *New York Times*. https://www.nytimes.com/2024/03/20/climate/anthropocene-vote-upheld.html.

8 Sustainability

No Other Choice

Environmental sustainability is frequently defined as creating a society that lives within the earth's limits without degrading the planet for future generations. This chapter explores the sustainability concept including introducing you to the United Nations' 17 Sustainable Development goals. Like it or not, sustainability is not a choice because natural processes have hard limits. We will either find and plan for a sustainable transition or sustainability will be forced on us. Our global economic system is a subsystem of the ecosphere and we are therefore ultimately dependent on earth's health. How do we go about making economic systems that respect earth's limits? We'll look at the field of ecological economics that seeks to do just that – find a bridge between these two homes – nature's home (ecology) and our fiscal home (economics). Eco, after all, is the Latin word root for home. Individual change will not be enough. Systemic change is needed. So, we'll also look at how systems work and what makes them change. Details matter in designing sustainable systems – we'll look at two case studies. One that examines the global trade in invertebrates (primarily butterflies) and its potential role in conserving this diverse group. A second that covers China's recent establishment of a national park that both protects biodiversity and supports traditional cultivation of tea by local communities.

Nature Owns the Stadium

My dad was somewhat compulsive about collecting stray aluminum cans. Maybe it was because he was a geologist, "there is only a set amount of aluminum on the planet so it's one of the most important materials to recycle." This would have been okay if he didn't insist on picking up cans (along with trash) on his daily jog. This also would have been okay if he had a normal dad jogging outfit. He did not. My dad's jogging attire varied depending on daily temperature and could include a multi-colored fleece beanie hat, a Mexican poncho, mismatched T-shirt, frayed jean shorts, long socks, and of course good running shoes. "Good shoes are important for healthy feet." I now look back fondly on this eclectic ensemble, but as a teenager I was mortified. My friends would see my dad out for a jog and

DOI: 10.4324/9781003725404-8

look over at me wide-eyed and say, "Is that your father?!" Having a face that tells the truth even if my brain wants to lie, all I could do was mumble "yes" as my cheeks flushed pink. My dad was right about aluminum though and, to his credit, also right about fast fashion.

A sustainable world is hard to imagine but imagine it we must. Despite the challenge, this work is critical. It provides motivation and builds community. It reminds us what we are working for. Sometimes I ask freshmen what we can do to solve the seemingly intractable environmental challenges we face. Typically, after an excruciatingly long pause, a student gunning for an A raises their hand, "Recycle?" The look on their face says it all. They know any individual action they can take is like spitting in the wind. We need to shift a system, move a whole darn planet. We've created this unsustainable path, systems nested within systems, that has our planet speeding towards an increasingly undesirable state. How do we find those levers of change? Cloth bags and reusable mugs may be necessary, but they are clearly not sufficient. Climate scientist Katherine Hayhoe says our voice is our most important tool in affecting change. How will you use yours?

There are many competing definitions for the term **sustainability** and it has a similar history to some of the other "young" terms and phrases we've already discussed. Similar to words such as biodiversity, the word sustainability came of age in the 1980s – as people grappled with the rising environmental challenges brought on by human activities. Sustainability can be broadly defined as having industries, economies, cultures, and lifestyles such that we live within the earth's limits and leave the planet healthy for generations to come. Because of our planet's finite nature and the presence of natural constraints, sustainability will eventually happen whether we want it to or not. It's important to acknowledge that sustainability is not a choice. It will happen either by design or by default or, most likely, some combination of the two. As Andres Edwards writes, "Nature not only bats last [it] in fact owns the stadium" (Edwards, 2010, p. 7). Indeed. This might seem counterintuitive given that many current lifestyles and economies are not even close to sustainable. Why can't we just continue with business as usual? It's worked so far. The simple answer is that the pantry will eventually be bare. Yes, we can live wildly unsustainably in the short term (and we do, some of us much more than others) but going beyond earth's limits is not a long-term option. I'm reminded of the joke "My plan is to live forever. So far, so good." Eventually the keg for this global party will run dry. New technologies can and will buy us time and sometimes they even change the overall dynamics but, a basic truth persists: if we use resources faster than they replenish themselves, we are living on borrowed time. (The Haber-Bosch process for making liquid fertilizer is an example of how a new technology can dramatically alter dynamics. Development of this technology in the 1950s allowed for a huge increase in crop yields globally.) So yes, new technologies can and do sometimes buy us time. But eventually we will not be able to escape the math. The length of time it takes for things to renew themselves, if indeed they even can renew themselves, is incredibly important. A large aspect of sustainability is speed and its opposite – *slowing down*. If we are looking for a sustainable approach, we need to account for the time it takes ecosystems to do all manner of things: replenish fresh drinking water, grow new trees, make new fish, and so forth. The wonderful news is we live on a planet that provides us with an astonishing array of goods and services. A few fish can make even more fish given time, the right habitat, and soft, romantic music. All we have to do is leave them alone. But with more than 8 billion people on the planet, and levels of consumption per person increasing, that simple act becomes harder and harder.

The myriad gifts the earth provides us – **natural resources** – differ in their physical and biological natures; these differences include how and even if they can replenish themselves. We need to be mindful of these characteristics when we contemplate how our economies use different items. The material goods of the planet are often divided into two broad groups: **non-renewable** versus **renewable resources**. (We'll talk later about this terminology and whether the word resources itself is problematic – how differently we might think about our planet if we used the word gift.[1]) But for now we'll stick with using the term resources and return to our courting fish which would be an example of a renewable resource because fish can breed and produce new baby fish. Given the right conditions, the fish population, the total number of fish in any given area, will grow in number. Unlike this reproductive "magic" that living things accomplish, many of the goods that run our local and global economies and dominate our lives are made from nonrenewable resources. Nonrenewable resources have a fixed supply on the planet. They cannot make more of themselves. Sometimes non-renewables can be reused but reuse is a very different thing from reproduction. Non-renewables *cannot* grow in total amount. In addition to not increasing in amount, many non-renewables are not reusable, we only get one shot. Once they're used they change form to no longer be useful. Fossil fuels would be an example of this type of nonrenewable resource which is also non-recyclable. In contrast, aluminum is an example of a nonrenewable resource that is recyclable. Although there is only a finite amount of aluminum on earth, it can be reused multiple times. This is not the case for fossil fuels. Once combusted, the fuel's chemical energy has been transformed into kinetic energy, heat, carbon dioxide, and only waste byproducts remain. One shot and with every shot there is less and less on the planet.

It will come as a shock to no one that building a global economy around one shot resources is not a fabulous long-term plan, but that is exactly what we have done since the Industrial Revolution. The transition away from fossil fuels as an energy source is critical not only because of the environmental damage these fuels cause, but also because these fuels (oil, natural gas, coal) are nonrenewable *and* non-recyclable. It's just a bad plan to run your economy using these energy sources. Compare these resources to renewable energy sources such as geothermal heat, solar, or wind. Some of these renewables are available in such an abundant supply that they are in a subclass of renewables called perpetual resources. The sun and the wind would be examples. We don't deplete the supply of wind or sun rays when we capture their energy. They are that abundant. As you might imagine, perpetual resources are the very best bet for long-term, sustainable energy production. Energy from these perpetual resources is also now often cheaper than from fossil fuels, a win-win, and people are working on innovative ways to store energy produced so it is available even on a cloudy day with no wind.[2]

Part of what sustainability requires, therefore, is shifting away from nonrenewable to renewable resources and, whenever feasible, using perpetual resources as the first choice since their supply is so great we don't need to worry about depletion. It also means using nonrenewable resources very carefully given their finite, nonrenewable nature. To harness the energy of the sun, geothermal heat, or wind, we use technology (solar panels, wind turbines) that is currently made up from nonrenewable materials including metals. Although metals are a nonrenewable resource, unlike fossil fuels they are **recyclable**. We can melt metals down, reshape them, and reuse them. Aluminum cans for example, melt, reshape, reuse. It is folly to not reuse and recycle metal because there is only a set amount of any given metal on earth, and we know we will need metals for future machines. Ideally, getting these required metals won't involve shifting through all our old landfills to find them. Better to initially use these gifts cautiously, carefully, and wisely. Part of sustainability therefore means creating items such that when they no longer serve one function, their parts can easily be reused to make something else, and when this second item breaks, its parts can be used to make something else, and when this third item breaks, and so on .... This is the idea behind a **circular economy**, creating products so as to minimize waste and maximize reuse. Again, we get back to the importance of time while introducing a new element – the importance of design. Designing items for long-term use and reuse. If we have taken a precious resource (gift) from the earth, it makes sense to get the maximum use we can from it. Even if more of a given metal can still be mined, reuse decreases the need to mine, therefore opening up land to other more environmentally-friendly uses. So, one aspect of sustainability is using renewable resources at a slow rate, a *rate* equal to or, ideally, much less than the rate that they can replenish themselves. Nonrenewable resources never replenish themselves so we need to be very cautious indeed in how we use them. Luckily, some non-renewables such as minerals and metals can be reused. They don't grow in amount but they can be repurposed again and again. To do this though, we need to design products that make such reuse easy. And just having these easily reusable products per se will not be enough; we also need to be able to complete the cycle by having industries that process products for reuse. This means having efficient methods and industries for collection, redistribution, and returning products back into basic materials that can then be supplied to manufacturers. The whole system requires readjustment because, unfortunately, our current system is too often a one-way ticket to the landfill (**linear economy**) and that is not sustainable.

How do we determine how far our current global society's consumption is from that of a sustainable society? One method is to measure a country or person's **ecological footprint**. Most of us are familiar with the concept of a **carbon footprint** – which estimates how much atmospheric carbon a person or country creates in a given period of time. An ecological footprint contains one's carbon footprint but expands the calculation to also include impacts beyond carbon emissions such as use of fresh water, deforestation, waste production, and pollution. How on earth can such disparate types of impacts be lumped together and quantified? This is not a trivial problem because we have to figure out how to get various impacts into similar units to handle them collectively mathematically. Anytime we measure something, that associated number should have some type of units associated with it. If you tell someone you are five-five, they assume you mean five feet and five inches not five meters and five centimeters. Units matter. Unfortunately, people sometimes try to mislead with numbers and lack of units is a red flag. If something looks like math, even going so far as to have a formula, *but* it doesn't have units, we can be pretty sure we are dealing with an attempt to make something look more quantitative and rigorous than it is; Carl Bergstrom calls misleading folks with phony formulas **mathiness** (Bergstrom and West, 2021).[3] Words x no units = It should have been a sentence. Ecological footprints pass this test; they quantify environmental impact using area – specifically total area of land and water (hectares or acres). It can be hard to convert all of our environmental effects into area but that is what these analyses do in order to be able to quantify environmental impact. For example, how much forested land is needed to provide the quantity of wood a country uses and to absorb its atmospheric carbon production; how big a body of water is needed to provide all of the freshwater; and so forth. We affect our environment both through consumption and through waste so this analysis also quantifies the land needed to process waste; basically, determining how big a landfill you need for all that crap. You can calculate a personal footprint using the Global Footprint Network's calculator (Global Footprint Network).

Not surprisingly, when we do an ecological footprint analysis we find there are dramatic differences in calculated footprints among individuals and among countries. If you have heard that we would need six planets if everyone on earth had the United States' levels of consumption, this is where the numbers for those calculations come from. Again, we get back to rates – our current rates of consumption use resources in large amounts too quickly, at a rate that can't be sustained over long time periods. We are depleting the earth's stocks. Overfishing is one example of taking more than the earth can give. If fisheries take too many fish from a healthy population, there will not be enough fish left to reproduce. The collapse of the Atlantic Cod fishery in the early 1990s is a classic example of overfishing, which lead to both meager population sizes of cod and collapse of associated livelihoods in the Northeastern United States and eastern Canada. After a peak catch of 800,000 tons in the 1960s, cod numbers started to decline precipitously. Mathematical models suggest that, if the cautionary data from the 1970s had been used to rein in fishing at that point in time, a sustainable fishery catch of 200,000 tons annually could have been achieved (Pope, 2021). Unfortunately, we can look back now on this as the sustainable path not taken. Heavy fishing continued despite warnings resulting in a complete collapse of both the cod population and people's livelihoods. Atlantic cod still swim in the Atlantic Ocean; they have not gone extinct, but they were severely overfished to the point that regional economies suffered. The Atlantic Cod population has not recovered. Overfishing is hard on the fish and hard on the angler. This historic example demonstrates why a plentiful fish harvest today may be a short-term boom that creates a long-term bust. Because of globalization and global markets, people's regional landscapes can get used and degraded in this way. Part of sustainability then is understanding the players in your local community and working to make a community that can sustain what your community values over the long term: businesses, people, and the natural environment. This is where the idea of **glocal** comes from: think globally, act locally. What might a local, sustainable community look like where you live? If global industries are a big part of your local economy, how are they interacting with local players and is this interaction good for the long term? How diverse is your community in terms of businesses and biodiversity? We'll see later that diversity is a key attribute to system resilience. Variety is good in a constantly changing world.

Revisiting a Classic Tale

We live at a critical time when human activities are using the resources (gifts) of the planet unsustainably. As we've seen, this means using resources too quickly so the given resource doesn't have adequate time to recover and replenish. It is critically important to distinguish unsustainable *rates* of a given activity from the activity in and of itself. The cod fishery could have been managed to produce a large annual sustainable catch of cod. A decreased catch in the 1970s and early 1980s would have paid off long term. To reinforce this point, that sustainability is all about rates; let's look at a fictional example. Many of us are familiar with the environmental tale *The Lorax* by Dr. Seuss (1971). In this story, the Onceler comes to the colorful truffala forest, sets up his shop, and starts cutting down trufflala trees to make thneeds out of the truffala tufts (if you haven't read the book, thneeds are like pink fuzzy jammies). The orange, fluffy Lorax creature pops out of the stump of the first cut tree and immediately begins lecturing the Onceler, shaking his finger. It is wrong to cut the trees, wrong to pollute the water, wrong to destroy habitat for the Barbaloots (adorable bear-like creatures) and so on. In this well-known fable, the cutting of even one tree is portrayed as morally wrong and environmentally problematic. The story casts environmental issues as a righteous morality play of us against them (the Loraxes vs the Oncelers), and the reader is clearly encouraged to take the side of the Lorax ("unless someone like you cares a whole awful lot"). People and their use of the land are presented as the problem. But this casting of people as outside of a healthy natural system does a disservice to the many peoples of the earth who have lived sustainably for millennia. Indigenous groups of the Amazon forest come to mind, or Native American groups such as the Menominee in Wisconsin. The Menominee have harvested the trees in their forests in such a way as to protect their forest's vibrant biodiversity while also extracting lumber. In contrast, The Lorax sets the stage for a political fight – you need to care a lot and stand up against the Onceler. Instead of posing environmental solutions as a political conflict, what if the question posed was how to live in this place (the magnificent truffala forest) in such a way that it continues to provide across generations – for the sweet Barbaloots, for the fish, for the trees, *and* for the Onceler and his family. This seems like a legitimate problem to solve – what does a sustainable thneed industry look like? How can the forest and water be kept healthy and clean? What exactly are the waste products of thneed production? How much habitat should be protected to keep Barbaloot populations healthy? *The Lorax* is a colonial story – the story of a colonizer who comes and takes and destroys and then packs up and leaves once the land and water is degraded. In this respect, it does have a story to tell but it isn't a sustainable story. In contrast to the cautionary colonial tale, a sustainable story is a story of place. A story of studying trees and fish and bears to learn and respect their needs. A story where thneeds are made well and last but can become quilts which at the end of their use can be composted to make new soil to germinate truffala seeds. A sustainable story is about knowing and living in one place. Living as if we are part of nature and don't have anywhere else we need to get to.

Inevitably, when we reflect on sustainable societies we move away from just the ecology of the system (production of wood or freshwater or fish) to thinking about the people interacting with nature in these systems – cultures, societies, and economies. How are the land and waters viewed? Do people stay in one place or are they restless and on the move? Because many of us interact with a globalized market, interactions get very complicated very quickly. Our local lives and economies are subsystems within this enormous global system. We'll get back to why the structure of our global economy tends to degrade ecosystems when we discuss how classic economic systems interact with nature. But first, let's explore the variation we see among societies in their environmental impacts. When looked at as one global number, ecological footprint analysis can give the impression that all people on the planet are living in a reckless and unsustainable manner. Both footprint analysis and the Lorax fable make us see people as the problem – we arrive, we consume, we degrade, we leave for greener pastures. And yes, if you look across all 8 billion of us, the math is irrefutable: the center cannot hold. But let's think about how we go about calculating a global ecological footprint; we are averaging, combining all countries (and all people). This analysis combines people with widely divergent income and consumption levels, not to mention all of the various global cultures. Scaled globally, indeed we need six planets to feed the consumption patterns of the United States (average US consumption multiplied by 8+ billion). So, of course, part of sustainability means reining in consumption patterns of high-income countries and within those countries, specifically, reining in consumption of the wealthiest individuals who have the largest footprints. For those of us who have grown up in a consumer, capitalist culture envisioning a sustainable society can be challenging. Many of us are stuck in a paradigm that thinks of high consumption levels as a given. However, there are societies and cultures that have shown themselves to be sustainable. As the saying goes, the proof is in the pudding. These peoples have lived sustainably in their landscapes for millennia. It's useful to examine current indigenous societies because they exemplify living in place. To repeat myself, a sustainable story is about living in one place and knowing that place. Living as if we are part of nature and don't have anywhere else we need to get to.

There are indigenous societies across the globe and often these groups are involved with managing local landscapes and their associated biodiversity. Sadly, this is not always the case due to the violent history of genocide and displacement of indigenous people from their lands such as occurred in the United States. In some locations, indigenous groups still live on their lands but have yet to secure tenure rights so ownership may not be recognized by local governments. Such land tenure rights for indigenous groups are a critical step in our move towards global sustainability given that these cultures and peoples are models of sustainability (Reytar, 2024). We saw in Chapter 7 that loss of the earth's genetic diversity was happening so fast that it places our planet at the highest risk levels (Figure 7.1). One of the eight principles of this book is "you are part of nature." Indigenous societies more than others have developed in deep acknowledgment of this reciprocal relationship between people and their local ecosystems (Kimmerer, 2011). Robin Kimmerer writes, "in the indigenous view, an authentic landscape incorporates human participation in ecological flourishing." In the Amazon forest, it is indigenous territories that are showing the highest resilience in the face of logging and wildfire (Box 7.1). What Western societies label as natural resources are viewed in indigenous societies as gifts and a gift should be reciprocated. One group that is working to sustain indigenous people and their lands is the Indigenous Conservation and Community Areas Consortium (ICCA Consortium, 2021). This group refers to the lands managed by indigenous groups as "territories of life." Unfortunately, because of the dominant Western perspective that people are both outside of nature and damaging to nature, indigenous use of ecosystems has sometimes been deterred and even criminalized. This stops practices that are dually important for indigenous people and ecosystem health. For example, in the Philippines when the

Tinuy-an Falls Protected Landscape was set up in lands traditionally used by the Manobo, individuals were charged with illegal occupation and were told they could no longer use the land (ICCA Consortium, 2021).

When we examine indigenous approaches to the land we find several practices which help us in reflecting on sustainability. One attribute would be the smaller scale of the activity, subsistence-level use or small-scale markets which allow harvesting resources (receiving gifts) at a slower rate. Often resource-use philosophy is culturally embedded. The Menominee of Wisconsin have been cutting trees for centuries with no adverse effect on their forests. Research shows that the plant diversity in the Menominee forests does not differ between logged and unlogged sites. This diversity is also very high, more suggestive of an old growth forest than of a forest that has been providing timber for centuries (DeJoode, 2003). The Menominee call themselves the Forest Keepers and sell certified sustainable lumber. The harvesting methods are based on tradition as the following quote reveals:

> Start with the rising sun and work toward the setting sun, but take only the mature trees, the sick trees, and the trees that have fallen. When you reach the end of the reservation, turn and cut from the setting sun to the rising sun and the trees will last forever.
>
> Chief Oshkosh (Menominee Tribal Enterprises)

For many indigenous peoples, certain species also have strong cultural significance serving as both "cultural and biological keystone species" (Kimmerer, 2011). For the Anishinabe, this would include wild rice, for other peoples – salmon. There is an understanding in indigenous cultures of the link between human and ecosystem health. For example, the Onodaga Nation is restoring a highly polluted lake on their lands in North America and they do not view the restoration as complete until the fish are healthy enough so that the people can once again safely eat them. This view of the human relationship to nature is both one of belonging and one of having a responsibility to tend local ecosystems, serving as their steward. If this is done well, nature will continue to provide. Unfortunately, indigenous communities and their lands are under threat. The 2021 report by ICCA Consortium examined 17 case studies for indigenous lands and found threats from extractive industries in all but one case study. Frustratingly, these communities can also be threatened by mainstream conservation efforts which may take advantage of practices that protected biodiversity by acquiring the land and then displacing the land's people. There is a long history of such displacement – in the United States native people were displaced from their lands to establish two of our flagship national parks, Yellowstone and Yosemite. But there is a better way, integrating people's livelihoods and conservation is not only possible; it is happening. For example, in China a balance is being found between protection of biodiversity and protection of centuries of tea cultivation (Box 8.1).

Box 8.1 Case Study – Tea cultivation and Wuyishan National Park

The human history of the Wuyi mountains in southeast China runs deep: archeological evidence suggests people have lived in the region for 4000 years. Cultivation of tea in this area dates back to the Tang Dynasty (618–907 AD). The mountains' subtropical forest is also rich in biodiversity which led China to pilot establishing a national park in the area in 2015, named Wuyishan National Park (WNP). This young park covers 982 km^2 and encompasses three previously distinct protected areas. It is designated by the United Nations as both a World Biosphere reserve and a World Cultural and Natural heritage site. Prior to the park's designation, several reforms were put in place over time to increase forest protection. These included a ban on commercial logging in 2008 and a ban on expansion of tea orchards in 2011. A study done in 2022 looked at tea growers' perceptions of the park to understand the challenges this area faces as a socioecological system (He and Su, 2022; Figure 8.1). Three thousand people live within the park's boundary and approximately 20,000 live within two kilometers. The forest and tea production are viewed as integrated by tea growers, making up key components of what is simply called "the *shan chang*" (the mountain). But there are tensions in the system as mul-

Figure 8.1 Tree grower in rural China
Source: Jala1V (2008).

tiple goals are sought. Tea is cultivated to sell on the market with income from tea production positively related to family size, tea orchard area, and number of family members working as growers. Because of the 2011 ban on orchard expansion, ownership of orchards within the park is stable. Although most of the tea growers within the park did not view park conservation as affecting their livelihoods, 43 percent still viewed conservation as disruptive. Comments associated with this negative view included complaints about regulations against forest use, restriction of soil amendments, and restrictions regarding roads. There was also concern that tea orchards would be lost to the park and associated with this, concerns regarding compensation if such losses were to occur. When asked about **ecosystem services** they valued, tea growers ranked tea cultivation as the top provisioning service followed by provision of fresh water. Ecotourism was ranked as the highest cultural service and air purification as the top regulating service. Those involved in tea cultivation understood that growing tea provided value outside of the income derived including increases in physical and mental health, passing on of cultural heritage, and social stability (He and Su, 2022).

For now, China appears to be finding a balance allowing preservation of both its natural and cultural heritage (Figure 8.1). This approach is not unique to this specific park. According to Wang and others (2023), the management plans of all of China's parks seek to maintain some traditional agriculture. The goal is to make sure that biodiversity protection goes hand in hand with respecting the rights of local residents and their need for a robust local economy. In WNP, policies which restricted both logging and the total amount of area used for tea cultivation created boundaries for human use which, at least at present, appear to have prevented overuse of this awe-inspiring subtropical forest.

The Problem with Classic Economics

Learning from indigenous cultures can help us as we seek to develop a more balanced approach to our use of land and water. However, this approach can seem distant for those of us who are outside of these cultures. For indigenous and nonindigenous alike, many of us get most of our goods from a global market-place. In this global economic system, the dominant way we interact with land and water is indirectly through our patterns of consumption and associated production of waste. Why has global economics lead to such degradation of earth's ecosystems? Degradation to the point where the whole earth system is at risk? Some would say the answer is obvious, economic and ecological goals are, by nature, in opposition to one another, the well-worn jobs versus the environment trope. However, both ecology and economy share a common word root – **eco** – meaning home. Ecological economists are a group of academics working to reconcile these two sometimes warring houses: our financial home with our natural home. Exploring the field of ecological economics helps us understand why our current global economy has been so destructive to the environment. Peter Brown and Peter Timmerman (2015) describe classic economics as a theory that will not die despite, "grinding itself into the physical face of the planet." A systems view would tell us that some critical **feedback** must be missing. What stops economic systems from destroying the very planet they depend on? When we model systems, we always leave something out. This isn't a flaw of the approach; it is a necessary step that allows a model to focus on a given problem or question. In the case of classic economic theory, however, the earth itself is often absent. Such theory often disregards environmental constraints, tending to view items as substitutable rather than unique and necessary parts to a functioning system. In a classic view, the environment is even treated as an infinite place from which resources, raw materials, come and into which wastes go. This assumption cannot hold forever. One key difference between classic economics and **ecological economics** is the latter's insistence that the economy be treated as a subsystem of the global environment. A healthy economy is a system embedded within the ecosphere. The embedded nature of our global economy is undeniably true. If the sun disappears or all the plants go on strike and chlorophyll abruptly stops working, the global economy is definitely going to tank. Follow me for more insightful economic tips! Because many societies have taken ecosystem services such as soil formation, water cycling, growth of plants (productivity) for granted, the impact we have on the earth when products are created and used is not factored into pricing. This can lead to what economists call **market failures**. The climate crisis is a classic example of a market failure where the market is obviously not maximizing what is good for society (Bowen, Dietz, and Hicks, 2014). We go along merrily pumping out greenhouse gases, passing planetary boundaries with no feedbacks to the economy or to the consumption of fossil fuels. Economists suggest such market failures happen because the damages of using a given product (in our example, fossil fuel use) are not accounted for in the product's pricing.[4] Pricing failures occur because of what economists call **economic externalities**. An externality is an external cost (or benefit) that arises from production or consumption of a good or service. This cost or benefit is not accounted for in the item's price. Going back to our climate crisis example, carbon pollution would be a major cost (negative externality) associated with consumption of fossil fuels. It is not accounted for in the price – we don't pay the cost of absorbing all the carbon dioxide we produce when we buy gas. Externalities can also be defined as impacts (positive or negative) on a third party that was not involved in any given economic transaction. Again, climate change provides many examples of this – for example island nations would be such third parties, struggling with the impacts of sea-level change caused by emissions from larger and more powerful nations. When we think about environmental impacts of products (externalities), most of what pops to mind are negative effects but externalities do not have to be negative they can also be positive. If my city plants pollinator plants in local parks and my neighbor's honeybees benefit from the plant's nectar, this is a positive externality not included in the price of the plant. If I am vaccinated so a virus doesn't spread in my community due to herd immunity, this is a positive externality. If a teacher teaches someone to read and they then can teach their children to read at no cost, this is a positive externality. Goods and services that produce positive externalities are a fabulous deal. You literally get more than you paid for! It makes sense to find ways to subsidize and increase such activities. This is one reason we should plant more trees. The cost of planting the tree is minimal next to the benefits it produces across its lifetime. But what about products with negative externalities? What do economists suggest can be done to minimize the negative outcomes produced by them?

Economists would solve market failures with a policy solution which in economic jargon is referred to as internalizing the externality. If you know some product has a negative effect, you need a policy that makes the price reflect that external cost. Since we successfully developed policy for the acid rain challenge in the 1970s and 80s, let's use that as a case study on how such policies can work. First, the basic science to get us up to speed on this particular externality. Up through the 1980s we burned a lot of sulfur-rich coal. This burning decreased the pH of rainwater substantially because emissions formed sulfuric acid, a very strong acid, in the atmosphere. Acidic rain is bad for all manner of reasons – damaging to lakes and their biodiversity, bad for soils (leaches soil nutrients), erodes statutes and buildings and more. How did we get the coal industry to change? Regulation. Specifically, in 1990 the Clean Air Act in the United States was amended to allow a **cap-and-trade** program for sulfur emissions. A cap was placed on total sulfur emissions with the cap set to decrease over time, coal plants had to stay below individual caps unless they purchased emissions from a coal plant that had "extra" emissions to trade. Specifically, plants with low emissions that were below their cap could sell the emissions difference. This policy provided economic incentives for coal plants to upgrade and decrease sulfur emissions quickly. Environmental policies such as cap-and-trade can internalize an economic externality and, by doing so, use the market itself to force desired changes. Why don't we just do this for all major negative externalities? One reason is politics, changing policies is hard work and can be unappealing particularly to industries that currently benefit from not having a proposed regulation. Industries tend to fight such policy changes. Another, associated reason, is that the price of certain items would be astronomical if we paid their "true price." Yes, it's true no one could buy these items at their true price, but the impact would also be severe because of our current dependence. We have built a system that is very addicted to having third parties, including future generations, pay the cost of externalities. Sovacool, Kim, and Yang (2021) analyzed the externalities associated with global electricity and transport. The estimated cost of these externalities is more than 26 trillion which is more than 28% of global gross national product, gulp. Their paper's abstract ends with this pithy sentence, "The fundamental policy question is whether we want global markets that manipulate the presence of externalities to their advantage, or a policy regime that attempts to internalize them." If we go with internalization, this must be done in a way to wean ourselves off products quickly but carefully rather than immediately impose a shock to current systems. The one thing we do know is that the more negative externalities can be internalized into prices and policies, the more nudges we are making to push our economic system away from its current damaging effects. The flipside, and no doubt easier piece to garner support for, is to increase activities that have positive environmental and social externalities. A lot of the activities we'll discuss in the next chapter fall into this category – rewilding, restoring, reusing, education and more. There may even be cases where international trade can support conservation. Box 8.2 examines the current trade in invertebrates from this perspective.

Box 8.2 Endless forms most beautiful and wonderful - Can global markets protect tropical butterflies?

"Endless forms most beautiful and wonderful," Darwin wrote this in the final sentence of *The Origin of Species* (Darwin, 1996, p. 396). These words are apt in reference to the incredible taxonomic diversity and gorgeous variation we find among the world's butterflies and moths. Charles Darwin is just one of many historic figures who was an avid butterfly collector, another was Winston Churchill. This hobby persists today (Figure 8.2) and collection is easier than

Figure 8.2 Glass collector's case filled with pinned butterflies and moths
Source: Brandt (n.d.).

ever with global trade and the ability to ship live butterfly pupae internationally. Shipment of pupae allows for a new type of market beyond collectors. Pupae are sent to butterfly houses (exhibits) across the world. These invertebrate zoos give visitors the experience of seeing live exotic adult butterflies in person. International trade in invertebrates is extensive and is dominated by trade in this group – the Lepidoptera (butterflies and moths; Lassaline et al. 2025). Wang and others (2023) tracked online sales of butterflies and found that 3,767 different butterfly species were sold with 96% of sales crossing national borders – primarily crossing from the Global South to the United States and Europe. Aesthetic ranking of a butterfly predicted its trade volume. On its face, such trade might appear to be a threat to wild butterflies. However, because of their fast life cycles, terrestrial invertebrates offer a unique opportunity for economics and conservation to potentially work hand in hand. Unlike vertebrate species such as mammals which produce few young and live long lives, invertebrates including insects have lots of young and live short lives. This fast life-cycle means that rearing butterfly pupae on a local farm requires much less effort and initial investment than raising larger animals. Local butterfly farming is therefore an easier lift for lower income and smaller rural communities. Tropical butterfly farming can offer disadvantaged communities a way to increase income while also helping to conserve local ecosystems and their invertebrate species. Case studies across multiple countries suggest sustainable butterfly rearing and trade is possible, particularly if markets avoid middlemen and directly link tropical butterfly farmers to consumers (Parsons, 1992).

Walking into the Butterfly Pavilion in Denver, Colorado, you immediately see flashes of color in every direction – brightly colored tropical butterflies flit through the air – going from plant to plant, occasionally landing on a wedge of orange to drink some juice. The air is so thick with butterflies, one staff member's job is to monitor a small room which is set between two sets of exit doors. People stand is this room while exiting while the staff scan the room to ensure no butterflies escape to the gift shop. "What makes all the butterflies come *here*?" asks an awe-struck child. Denver's Butterfly Pavilion, like all butterfly houses, is perhaps best described as an invertebrate zoo although most zoos don't let you walk right into the animal enclosures. Butterfly houses are immersive experiences for visitors – providing areas much like a botanical garden or aviary but with butterflies from around the world flying among the plants – if you're lucky one might take a rest by landing on your head. In addition to education and entertainment, these invertebrate-focused organizations support research and global conservation projects – providing a market for tropical butterfly farms. Increasingly such exhibits are not just in the Global North but are also found where the habitat and local farms for these tropical invertebrates are located. For example, the BBC Butterfly Center in Cambodia (n.d.) is in Cambodia's Siem Reap Province. Income from tourists who visit the center funds local research and conservation while also providing income to local communities. The Kipepeo project in Kenya (n.d.) also houses a local butterfly exhibit. This project started by selling just butterfly pupae but now also sells other invertebrates as well as silk cloth and honey. The strength of such projects is that the money from rearing butterfly pupae supports livelihoods in the communities where conservation needs to occur to protect global butterfly diversity. Having local communities benefit directly from biodiversity conservation is best practice for long-term nature protection – making sure that income from butterfly farming goes directly to those involved in butterfly rearing. With any enterprise there is always the risk of overexploitation, but these centers demonstrate how butterfly conservation and economic development can be a win-win scenario.

We will have to reconcile these two systems – the ecological and the economic. Meadows (2008) suggests that one way to cause systems to shift is to create a new paradigm, a new way of viewing the system, a new set of working assumptions. Kate Raworth (2017) argues for such a new mindset in her work on **doughnut economics**. She uses the planetary boundaries framework covered in the last chapter (Figure 7.1) but adds to it an inside circle which she labels "social foundation." The idea is that we need economies that provide social goods and it is very problematic to society if they fall below a certain level. However, we also need economies that don't exceed ecological limits (she calls this the ecological ceiling). The doughnut in doughnut economics emerges as the operating space that exists between the minimum level of activities that meet the world's social needs and the maximum level where the ecological limit is exceeded. Getting economies within this safe operating space is the goal.

It's All Interconnected: Systems Thinking

Our society and the natural environment are interwoven and interdependent. Since the 1980s the United Nation's (UN) reports have provided a framework for thinking about global environmental challenges including their societal components. One piece of this framing is the **3Es** – environment, economy, and equity. A sustainable future is envisioned as a society that has both a healthy environment and economy, as well as greater equity so that the benefits of environmental and economic health are shared and accessible. With these three broad objectives in mind, we can better understand the 17 narrower **sustainable development goals** articulated by the UN in 2016. Table 8.1 lists these goals. The UN reports also provide indicators along with each goal – aspects of the global system we can measure and monitor to track our progress. Many of the 17 goals are societal in nature such as decreasing poverty and hunger and increasing equity. The links between a healthy environment and human health and well-being are found in goals such as clean water and sanitation. Strong institutions are needed so we can both develop and successfully implement policies that address

Table 8.1 United Nations sustainable development goals

Goals 1–9	*Goals 10–17*
No poverty	Reduced inequalities
Zero hunger	Sustainable cities and communities
Good health and well-being	Responsible production and consumption
Quality education	Climate action
Gender equality	Life below water
Clean water and sanitation	Life on land
Affordable clean energy	Peace, justice, and strong institutions
Decent work and economic growth	Partnerships for the goals
Industry, innovation, and infrastructure	

Source: United Nations. Department of Economic and Social Affairs. (United Nations).

global environmental challenges such as biodiversity conservation and climate mitigation. Education is needed on all fronts so that there is a greater understanding of the complicated, interconnected nature of the challenges we face. There is also a need to continually redesign global systems so that we make progress on societal goals while accounting for environmental realities. Climate action and associated clean energy are, not surprisingly, two of the 17 sustainable development goals. The great news is that the price of renewable energy is now lower than that of fossil fuels, making it more feasible than ever to embrace the transition away from fossil fuels to a clean energy future. When the UN set these goals, they gave the world a 15-year time period to meet them: 2030 was the deadline. But creating goals is an easier task than shifting global systems so that the goals are met and we are still far from meeting these goals. Walter Leal Filho and others examined challenges in implementing the 17 goals and identified climate change, the Covid pandemic, and war as three major negative influences (Leal Filho et al., 2023). As we work to achieve these goals it's important we acknowledge their strong interdependence – conservation isn't possible without strong public health and strong institutions and communities. We will not get to where we wanted to be by 2030, but that doesn't mean we should stop trying. We must continue to learn from our failures and setbacks as we modify our approaches and continue to strive towards a healthier and more equitable world.

Ecosystems are complex systems that can behave in unpredictable ways, and our social and economic systems are subsystems within a global ecosystem. This creates a lot of unpredictability and uncertainty. If we hope to achieve the UN's sustainability goals, we need to understand how to design systems that meet our present needs but that increasingly account for impacts on both nature and future generations. Often environmental textbooks will devote large sections to "what you can do" as an individual: buy energy efficient appliances; support politicians that will support environmental policies; bike to work; and so forth. But what if it's not safe to bike to work? What if you can't afford new appliances? Individual actions can be thwarted by systemic barriers. For most of us, with the exception of the select few who hold great power, the greatest influence we can have is through work we do to change the system. Specifically, working towards shifting the systems we are part of towards a new system state. I could modify my personal behavior and consumption dramatically and it would not be a blip on the radar of the global state of things. This is where thinking of societies as systems can help. We need to redesign the systems and this is hard work indeed. How does one change the behavior of a global system? Dana Meadows spent her life asking just this question and she suggests a variety of system leverage points in a book I highly recommend everyone read.[5] The leverage that Meadows (2008) contends is most powerful might surprise you. It is not a change in policy or a change in goal, although these are strong leverage points she discusses and are ones we are familiar with. Clearly, as the discussion of cap-and-trade suggests, such systemic changes can be powerful levers moving the system in a better direction. Yes, we need clean energy policies, policies that push us to wean ourselves from plastics, regulations on pollutants. While I'm discussing policies, I'll make a push for a biodiversity and ecosystem services act that ties conservation to the services we want from ecosystems. Policy changes can and do push the needle. The Endangered Species Act saved many United States species from extinction. Policies and the laws associated with them are a powerful lever, but instead of aligning the compass needle a bit better with the direction we want to go, what if we got an entirely new map after realizing we had brought the wrong one with us? All this time spent looking at the terrain incorrectly.

Meadows' (2008) argues the strongest leverage point for creating change in systems is shifting societal assumptions – what she and others call **paradigm** shifts. Meadows defines paradigms as "shared societal agreements about the nature of reality." As an example, she gives the assumption that people can own land. In Western, capitalist societies the idea that land and property are one and the same is assumed because this is how land is treated within this social-economic system, as a commodity. Other cultures and societies, however, have viewed land, and water, much more expansively. For example, land can be viewed as a place: a place with deep cultural, ecological, and historical functions and meaning (Goeman, 2015). Misuana Goeman (2015) investigates land concepts in indigenous cultures and discusses this layered perspective. In this way of thinking, there is something off about viewing land as a simple commodity that can be easily bought or sold; that certain people can own and others can't. Other paradigms are also deeply rooted, even in environmental work. One of these would be an anthropocentric perspective – viewing the world and its "resources" from a human-use angle. Environmental texts call the goods and services provided

ecosystem services, that is, what ecosystems provide to people. In an anthropocentric paradigm, the arrows all point to the people at the center. What has this ecosystem done for *me* lately? But not all people on the earth share this paradigm and when we look to cultures that have proven themselves to be sustainable, we do not find a me-first, prideful, humans are the best better than all the rest, anthropocentric view. Instead, we find a paradigm of **reciprocity** (Kimmerer, 2013; Kimmerer 2024). Instead of resources, this worldview sees nature as a provider of gifts, and a gift requires acknowledgment and respect for the gift giver. A gift should be reciprocated. People, in these cultures, feel an obligation, a responsibility, to give to nature. People are nature's gardeners, or nature's stewards. We take care of nature and nature takes care of us. Some cultures, of course, would go even farther than ideas of reciprocity to view nature as sacred or divine. It is useful to contemplate such paradigms and reflect on how changing such a broad societal worldview would shift your own understanding of the world, your place in it, and what actions you might take. There are many indigenous scholars who discuss such paradigms. Dr. Robin Kimmerer is perhaps the best known in having brought these worldviews, including articulating concepts of reciprocity, to a broad audience in her books *Braiding sweetgrass* (2013) and *The serviceberry* (2024).

Changing our view of human-nature relationships to one of mutual support and obligation, continual gift exchanges, could therefore be one of the strongest leverage points to creating large-scale systemic change. The reason why paradigm shifts are powerful levers in systems is because they call into question the very reason for the whole system. If a society changes its underlying assumptions about the nature of reality, system redesign will follow. I am not saying system design is easy. It is in fact incredibly hard; social and economic systems, like ecosystems, have complex subsystems that create webs of unpredictability. To return once more to Meadows' work, she writes: "The future can't be predicted, but it can be envisioned and brought lovingly into being. Systems can't be controlled, but they can be designed and redesigned" (Meadows, 2008, p. 169). People who study systems have come to accept that it is impossible to predict all system behaviors or force a given future into being. However, we do have the power to imagine the future we desire and from there, design and redesign systems so that they get us closer to that desired state. Our final chapter will look at ways people are doing just that.

Conclusion

Ecosystems are place-specific. Some areas of the world are forested, others have deserts or grasslands. Increasingly, landscapes are dominated by urban or agroecosystems. Sustainable communities will look different depending on where we live, what cultures we are a part of, and what economies we interact with. A gift economy looks very different than a global market-place. In the next and final chapter, I am going to treat you to all the things that give me hope. One of these is the field of **reconciliation ecology**, an approach that seeks to bring nature stewardship and biodiversity conservation to human dominated landscapes. What might that look like? Let's end this chapter by taking a quick trip to Australia to see what one community is doing to protect a cherished mammal: the koala bear.

Barbaloots are fictional Dr. Seuss bears, but the koala "bear" (*Phascolarctos cinereus*) is both very adorable and very real (Figure 8.3). Koalas are unique in being the only living mammal in the taxonomic family *Phascolarctidae*. These mammals are specialists who eat and live in eucalpt trees a group that includes 812 tree species that are native to Australia. Unfortunately, many of these native trees are in decline.[6] As eucalpt groves have decreased, so have populations of koalas which call these groves home. Koalas are currently classified as vulnerable to extinction by the International Union for Conservation of Nature.[7] These cute, grey, fuzzy mammals are more at risk in some regions of Australia than in others. One high risk region is New South Wales. What might communities look like if they prioritized the needs of the koala? The Koala Beach Housing development in New South Wales gives us a glimpse. Two of the major threats to koalas that tend to arrive along with people are cars and dogs. Koala Beach is therefore a no-dog community and cars can only be driven at low speeds. In addition, all trees that are habitat for koalas are conserved and fences between houses are elevated so that koalas can crawl beneath them as they romp around

Figure 8.3 Koala bear (*Phascolarctos cinereus*) eating eucalyptus leaves
Source: Cuatrok77 (2014).

the landscape. This koala-centered, suburban design has benefited not only koalas but also other plants and animals. We see in this community some of the elements we've discussed in this chapter including **reciprocity** – understanding the needs of another species and prioritizing them. In this case the two reciprocating parties are people and koalas. This Australian community found a culturally significant, native species to construct itself around. (You can find out more about this koala loving community at https://www.savethekoala.com/our-work/koala-beach-housing-development/.) Is there a plant or animal native to where you live that could serve that role? A creature worth driving slow for? What might need to change if the place you call home focused on the needs of this species as well as people's needs? It is the task of our generation to not just envision positive futures but to roll up our sleeves and work to make them a reality. In many places around the world, this is already happening.

Notes

1 Some people dislike the term natural resources and suggest that this language itself prevents a sustainable perspective. They argue that the word treats the earth like a pie made for people to divide up, rather than like a living planet which we should respect and repay for the gifts it gives us. There is strength to this argument and we'll return to it when we explore cultures that have lived sustainably for millennia. Feel free to substitute the word gift for resources to see how that substitution influences your perspective.

2 One exciting development is the affordability of solar and wind energy and breakthroughs in energy storage mechanisms. Because the sun and wind vary in intensity over time, being able to store the energy they produce to use later is key as we expand this energy source. Will storage be as hydrogen gas stored underground? Possibly (Małachowska et al., 2022).

3 If you want to delve more deeply into how to detect mathiness, I highly recommend Bergstrom's book (Bergstrom and West, 2021). You will find some common environmental formulas don't pass the unit test – the IPAT "equation" for environmental impact fails the unit test.

4 Whether economics is the right system to use to look at sustainability is ripe for hearty discussion. For better or worse, however, we tend to understand and respond to pricing. To learn more about climate change as a market failure see Bowen, Dietz, and Hicks (2014).

5 Please, please, please read Donella Meadow's book on systems thinking (Meadows, 2008). If you want to change the world, and who doesn't, read this book. The more systems thinkers we have the more likely we will actually figure out a way out of the mess we've gotten ourselves into! Triple please book recommendation.

6 Eucalypt as a tree group was evaluated by the International Union for Conservation of Nature (IUCN) and 25% were found to be threatened with extinction including species in Australia which are a food source for koalas and a cultural keystone species for Australia's First Nation people (IUCN, 2019).

7 The IUCN (International Union for Conservation of Nature) assesses the extinction risk of species and gives them a rating based on the best available data. Assuming adequate data is available these ratings range from least concern (best) to extinct-in-the wild (second worst, one up from extinction). On this scale, vulnerable (the rating given to koalas) is better than endangered but worse than near threatened.

References

BBC Butterfly Centre (n.d.) Livelihood. https://www.angkorbutterfly.com/livelihood/.

Bergstrom, C. T. and West, J. D. (2021). *Calling bullshit: The art of skepticism in a data-driven world*. Random House Trade Paperbacks.

Bowen, A., Dietz, S., and Hicks, N. (2014, March 21). Why do economists describe climate change as a market failure?https://www.lse.ac.uk/granthaminstitute/explainers/why-do-economists-describe-climate-change-as-a-market-failure/.

Brandt, B. (n.d.). *Butterfly collector Kauri Mikkola 1967*. Wikimedia Commons. https://commons.wikimedia.org/wiki/File:Butterfly_collector_Kauri_Mikkola_1967_(JOKAV-N4A2H13–12).tif

Brown, P. G. and Timmerman, P. (Eds.). (2015). *Ecological economics for the Anthropocene: An emerging paradigm*. Columbia University Press.

Cuatrok77. (2014). Koala aggripé à une feuille d'eucalyptus, Queensland, 2014. *Wikimedia Commons*. https://commons.wikimedia.org/wiki/File:Koala_aggrip%C3%A9_%C3%A0_une_feuille_d%27eucalyptus,_Queensland,_2014.jpg

Darwin, C. (1996). *The origin of species*. 1859. Oxford University Press.

DeJoode, D. R. (2003). *Effects of logging disturbance on understory plant communities in temperate forests*. Thesis. University of Michigan.

Edwards, A. R. (2010). *Thriving beyond sustainability: Pathways to a resilient society*. New Society Publishers.

Global Footprint Network. (n.d.). What is your ecological footprint?https://www.footprintcalculator.org/.

Goeman, M. (2015). Land as life. *Native Studies Keywords, 1*(1), 71–89.

He, S. and Su, Y. (2022). Understanding residents' perceptions of the ecosystem to improve park–people relationships in Wuyishan National Park, China. *Land*, 11(4), 532. doi:10.3390/land11040532.

ICCA Consortium. (2021). *Territories of Life: 2021 Report*. ICCA Consortium: worldwide. Available at: report.territoriesoflife.org.

IUCN Press Release. (2019, Dec 10). Species recoveries bring hope amidst the biodiversity crisis: IUCN Red List. https://iucn.org/news/species/201912/species-recoveries-bring-hope-amidst-biodiversity-crisis-iucn-red-list.

Jala1V. (2008). Rural tea grower in the hills around Hangzhou, China. Wikimedia Commons. https://commons.wikimedia.org/wiki/File:Tea-grower-hangzhou.jpg.

Kimmerer, R. (2011). Restoration and reciprocity: the contributions of traditional ecological knowledge. In: D. Egan, E.E.Hjerpe, J. Abrams (Eds.) *Human dimensions of ecological restoration: Integrating science, nature, and culture*, 257–276. Island Press. doi:10.5822/978-1-61091-039-2_18.

Kimmerer, R. W. (2013). *Braiding sweetgrass: Indigenous wisdom, scientific knowledge and the teachings of plants*. Milkweed Editions.

Kimmerer, R. W. (2024). *The serviceberry: Abundance and reciprocity in the natural world*. Simon and Schuster.

Kipepeo Butterfly Project. (n.d.) Promoting nature conservation and community livelihoods. https://kipepeo.org/.

Lassaline, C. R., Toomes, A., Fagan-Jeffries, E., and Cassey, P. (2025). From forest floor to market door: The global terrestrial invertebrate trade. *Biological Conservation*, 308, 111266. doi:10.1016/j.biocon.2025.111266.

Leal Filho, W., Trevisan, L. V., Rampasso, I. S., Anholon, R., Dinis, M. A. P., Brandli, L. L. ... and Mazutti, J. (2023). When the alarm bells ring: Why the UN sustainable development goals may not be achieved by 2030. *Journal of Cleaner Production*, 407, 137108. doi:10.1016/j.jclepro.2023.137108.

Małachowska, A., Łukasik, N., Mioduska, J., and Gębicki, J. (2022). Hydrogen storage in geological formations – The potential of salt caverns. *Energies*, 15(14), 5038. doi:10.3390/en15145038.

Meadows, D. H. (2008). *Thinking in systems: A primer*. Sustainability Institute.

Menominee Tribal Enterprises. (n.d.). Forest Keepers. https://www.mtewood.com/.

Parsons, M. J. (1992). The butterfly farming and trading industry in the Indo-Australian region and its role in tropical forest conservation. *Tropical Lepidoptera Research*, 1–31.

Pope, A. (2021, Aug 31). Study of 500 years of cod catch data shows collapse could have been avoided. *Canadian Geographic*. https://canadiangeographic.ca/articles/study-of-500-years-of-cod-catch-data-shows-collapse-could-have-been-avoided/.

Raworth, K. (2017). Why it's time for Doughnut Economics. *IPPR Progressive Review*, 24(3), 216–222. doi:10.1111/newe.12058.

Reytar, K. (2024, Nov 22). Protecting biodiversity hinges on securing indigenous and community land rights. *World Resource Institute*. https://www.wri.org/insights/indigenous-and-local-community-land-rights-protect-biodiversity.

Seuss, T. (1971). *The Lorax*. Random House.

Sovacool, B., Kim, J., and Yang, M. (2021). The hidden costs of energy and mobility: A global meta-analysis and research synthesis of electricity and transport externalities. *Energy Research & Social Science*, 72. doi:10.1016/j.erss.2020.101885.

United Nations. (n.d.) The 17 Goals. https://sdgs.un.org/goals https://commons.wikimedia.org/wiki/File:Koala_aggrip%C3%A9_%C3%A0_une_feuille_d%27eucalyptus,_Queensland,_2014.jpg.

Wang, B., He, S., Min, Q., and Yang, X. (2023). Conflict or coexistence? Synergies between nature conservation and traditional tea industry development in Wuyishan National Park, China. *Frontiers in Sustainable Food Systems*, 6, 991847. doi:10.3389/fsufs.2022.991847.

Wang, Z., Chan, W. P., Pham, N. T., Zeng, J., Pierce, N. E., Lohman, D. J., and Meng, W. (2023). One in five butterfly species sold online across borders. *Biological Conservation*, 283, 110092. doi:10.1016/j.biocon.2023.110092.

9 Conclusion: Hope in the Anthropocene

Solving interwoven global environmental problems is not something our species has had to grapple with before, saving our planet, maintaining the Holocene-like conditions that support human lives and livelihoods. I've heard speakers say that global change is our final exam. Looks like we've left things to the last minute and are going to have to pull an all-nighter if we want to pass. One way or another, change is coming and each of us has a role to play in making that change bring a better world to fruition. In this chapter, I discuss hope and activities that cultivate hope. Nature gives me hope so I cover hope from an ecologist's perspective: in the context of the science of ecological resilience and the emerging reality of **novel ecosystems**. We'll find that Mother Nature is one tough cookie and ecosystems can, and do, recover from the many onslaughts they face. People are also out there doing all kinds of amazing work. We'll look at the ecological and conservation science that links to these activities: ecological restoration, community-led conservation, and reconciliation ecology. Win-win scenarios exist where biodiversity protection can go hand-in-hand with urban design (urban ecosystems) and agriculture (agroecosystems). Because ecosystems are everywhere, ecosystem restoration can occur everywhere! For good reasons, the exact nature of successful endeavors will vary based on individual communities' histories, cultures, identities, and native biomes.

The task of addressing global environmental change is often both bewildering and overwhelming. Yet, every day, across the planet, teams of people are working to move the needle. Hope is an active verb. Despite extremely challenging odds, people roll up their sleeves, grab computers or shovels, put their shoulders to the wheel, and get to work. People in this line of work tend to get asked about hope a lot and it's worth thinking about what gives us hope. Environmental philosopher David Orr characterizes optimists as somewhat irritating: people who have the odds all in their favor. These folks have no reason not to be optimistic because the deck is, and has always been, stacked in their favor. Hope, according to Orr, is a different beast altogether. Hope is Emily Dickinson's "thing with feathers." Like the dinosaurs that flit from tree to tree (i.e. birds), hope can be hard to catch and can seem elusive but it is there. To quote Orr, "real hope, comes from doing the

DOI: 10.4324/9781003725404-9

things before us that need to be done in the spirit of thankfulness and celebration, without worrying about whether we will win or lose" (Orr, 2004). So, we cultivate hope, we plant hope through our actions. And there is a lot to do, so much so it can be overwhelming to find your small part. Marine biologist and climate activist, Dr. Ayana Johnson (2024) suggests that, given the wide variety of good work that needs doing, you should pick a task that you are both good at and that brings you joy.[1] So much needs to be done to get us back within earth's planetary boundaries, don't worry about doing it all (you can't), do one part. Do this work with others. Build community. We saw in the chapter on sustainability that part of what might "save the world" is literally changing our minds – viewing ourselves and our relationship to the planet as one of **reciprocity** (Kimmerer, 2011). Mutual aid would be another way we might phrase this. As more people embrace our role as nature's stewards or what Aldo Leopold would call a "plain member or citizen" of the land community, more will get done. It is past time to lean into indigenous paradigms of reciprocity[2] towards nature which, although not a new perspective for all, is new to many. Leopold's **land ethic** similarly embraced this idea of humanity, calling out our species for our pride and hubris. Leopold and others point us toward an ethic of responsibility towards other members of the natural community. Specifically, Leopold's land ethic is often summarized using this one quote,

> A thing is right only when it tends to preserve the integrity, stability, and beauty of the community and the community includes the soil, waters, flora, and fauna, *as well as people*.[3]
>
> (emphasis mine, Leopold, 1947, p. 1)

If we have a responsibility towards nature, if we are nature's gardeners or stewards, embracing nature across all landscapes and working to improve its health, our health, becomes imperative. Landscapes and waterways are interconnected and degradation in one part of the land escapes to other places. Pollution that comes off a farm field or road will be pulled by rainwater through ditches and sewers to degrade local streams and lakes. A farm and a stream therefore can't be treated as isolated entities. They are part of the land-water system. They are connected to and influence one another. Water runs downhill, and the cleaner we can make it, the better. If we want a local stream to have healthy fish and river otters, it means we need to think of making the water feeding into that location as clean as possible. I live in Central Illinois which is a region heavily affected by farming and urban development. Despite this, our small beleaguered local stream that cuts through my subdivision in Springfield is a favorite spot for local kids and my kid even saw an otter playing there once. Otters give me hope. So, when we think about United Nations sustainable development goals (see Table 8.1 for the list of goals), we should embrace these goals broadly. Using the example of the goal of clean water, we want clean water not just for people, but for the rest of the biological community, for the rest of nature. Clean water for the fish, for the otters, and for the kids who play in the stream.

Ecological Resilience

As an ecologist, one thing that continues to fuel my hope for the future are the many examples of nature's resilience. I am more worried about human societies (civilization) failing than of ecological collapse per se. Nature has been around a lot longer than us and has even weathered a mass extinction or two or five. (Recall from Chapter 6 that earth has undergone five mass extinctions in its history.) Nature has dealt with species changing the atmosphere dramatically and made it through. When photosynthesis evolved 3.4 billion years ago and the atmosphere shifted to high oxygen, it changed everything. Prior to this, anaerobic organisms could rule the earth (organisms that thrive without oxygen). Today, however, ancestors of these anaerobic bacteria that once dominated the planet are now banished to low oxygen environments. For example, deep down in the mud where oxygen can't penetrate. The earth has changed dramatically before including on a global scale. Of course, we return to the issue of time and the importance of rates; nature's past recoveries from mass extinctions or global atmospheric changes took millennia. Nature in this respect is incredibly patient and can spend this kind of time rebuilding biodiversity that is adapted to new conditions. Nature does fine with or without us. This gets back to that core principle from Chapter 1: nature is indifferent. She doesn't care about our hubristic little species or any other species for that matter. She works the way she works and will keep on going as physics, biology, ecology and evolution keep churning out changes. What is huge on the human Richter scale, is just a small nuisance to her.[4]

On a human time-scale, however, a 6th mass extinction or rapid climate shifts would be catastrophic to our species leaving generation after generation of people to deal with a depauperate earth affecting **ecosystem services** in unknown ways. We would have failed in our role as earth stewards if we allow this to happen on our watch, but it's not too late to avoid this bleak future. Dr. Peter Kareiva gives a talk where he shows a slide of a future city; one half is green and lush with streams along the roads while the other half shows the same city but grey and polluted – not a flower or bird in sight. He asks his audience, "which city would you want to live in?" He then proceeds to tell them that their answer reveals that they are all environmentalists whether they like the label or not. Although building back biodiversity from a mass extinction takes millennia, recovering species populations and restoring ecosystems, say a forest or coral reef, occurs over much shorter time frames, often 100 years or less. We just have to commit ourselves to this work across our lands and waters. Restoring the earth. Natural ecosystems can and do bounce back from incredible levels of destruction. In many places in the eastern United States, we can now stand in beautiful, diverse forests that are very much *not* old-growth forests. They are instead mature, second growth forests that regrew after widespread clear-cutting in the late 19th and early 20th century (Figure 9.1). Despite this history of extreme landscape scarring, Michigan is known today for its gorgeous, forested lands and recreational opportunities. Are these secondary forests the same as the old-growth ecosystems of the past? No, but they remain places of beauty, vitality, and biodiversity. They provide habitat to countless microbes, invertebrates, plants, and animals. This example is so heartening and the news gets better still because among the types of ecosystems found across our planet, forests have some of the longest recovery times simply because trees grow very slowly. It takes a good while for an acorn to become a mighty oak. Moving to our oceans, robust whale populations swim. Species such as the Humpback whale whose populations were once severely depleted by whaling now number in the tens of thousands (Mehta, n.d.).[5] Bald Eagles and other birds of prey are now abundant near lakes – no longer losing their eggs to the devastating effects of synthetic chemicals such as dichlorodiphenyltrichloroethane (DDT). It turns out such recoveries are the rule not the exception. Dr. Holly Jones and co-author (2009) reviewed rates of ecosystem recovery across a wide variety of ecosystems, both aquatic and terrestrial. Specifically, she looked at the extent ecosystems recovered from "agriculture, deforestation, eutrophication, invasive species, logging, mining, oil spills, overfishing, power plant[s], and trawling" (p. 91). Their review found that the majority of systems (72%) at least partially recovered in

Figure 9.1 Early 20th-century sawmill in Grayling, Michigan in what is now the Huron-Manistee National Forest
Source: US National Archives and Records Administration (1910).

mere decades and around a third of ecosystems made full recoveries over the periods they'd been studied (Jones and Schmitz, 2009). Nature can and does bounce back – the animals, the plants, and the ecosystem processes. It is best therefore to view ecosystems that have experienced degradation the way we view sick patients. These ecosystems require care. They can make a recovery in most cases, if we remove what made them sick in the first place, and do the necessary work to heal them. Sometimes, nature actually does some of this work for us. In 1992, Hurricane Andrew made landfall in Florida near Cape Florida State Park and took out 436 acres of trees. At first glance this might sound like horrible news, but there's a twist. The trees that were killed were nonnative Australian pines that had taken over the park along with other **invasive species**. The hurricane was thus a boon for park managers – who seized the opportunity provided by the hurricane's damage to do a large-scale restoration of native plants and animals. This was done with heavy involvement of school groups and community members (Westervelt, 2011). Ecosystems restored; community built.

Preserving the earth's **genetic diversity** is critically important to maintaining nature's power of resilience. As we saw in Chapter 7, the global rate of genetic diversity loss is of grave concern to those who study planetary boundary conditions (Figure 7.1). One reason it is so important to preserve the rich diversity of species and the rich genetic diversity within species is because this genetic variation provides a safety net for future ecosystems. High levels of diversity allow nature to be resilient by giving ecosystems lots of future options. We don't know what future environments will be like, but if our ecosystems are chock-full of genetic diversity, there is a greater chance that they contain varieties that will do well in those future environments. Genetic diversity gives natural systems multiple potential evolutionary pathways. *Evolution is a great reason for hope!* Across the planet, species are adapting to the changed environments we've created. For example, house finches in Arizona now have bigger, deeper beaks. Why? They've adapted to suburban environments which have lots of sunflower seed in bird feeders (Badyaev et al., 2008).[6] Coral reefs provide another example of the importance of genetic diversity. Reefs are at high risk as our oceans heat up because of their sensitivity to water temperature. However, it turns out that heat tolerance in corals has a strong genetic component. If coral from naturally warmer reefs are bred with individuals from cooler reefs, their offspring have higher survival when put in hot water. In addition, even without this genetic boost from other populations, there seem to be some individuals in the cooler reef coral populations that already have high heat tolerance (Howells et al., 2021). These heat-tolerant genotypes would be expected to be favored, having higher survival and reproduction, as the waters of the reef warm (Figure 9.2). As an ecologist, when I see studies such as this all I can think is, "well, thank God for evolution!" Yes, evolution can sometimes suck, but it can also at times be a godsend (back to Chapter 1

Figure 9.2 Coral propagation for a restoration project
Source: UVI (2022)

principles). There is also a role for us here, as nature's stewards, that goes beyond simply conserving genetic diversity. We are smart little primates and can grow heat-tolerant genotypes in labs. We can and we already do! These resilient strains can be introduced to sick reefs and hopefully bolster their resilience even more. In Mexico, conservationists are even 3D printing calcium carbonate reef structures to support ocean biodiversity (Ramirez, 2025)! Building nature together.

Generation Restoration

The United Nations declared 2021–2030 to be the "UN Decade of Ecosystem Restoration."[7] I've defined ecosystems as nature's unit (Chapter 5) and argued that ecosystems are everywhere (one of the eight guiding principles), which means that for our purposes, ecosystem restoration can be interpreted broadly to mean *enhancing the ecological health of all areas of the earth*! Let's get to it! What does it mean to restore the health of earth's myriad ecosystems? Let's look at a textbook definition of ecosystem restoration; the Society for Ecological Restoration (SER, 2002) historically defined **ecosystem restoration** as "the process of assisting the recovery of an ecosystem that has been degraded, damaged, or destroyed" (p. 2). Given how we've thought of ecosystems in this book, this definition can work for our purposes although it might encompass a lot more than this academic society would think! SER's definition doesn't constrain us to thinking of ecosystems solely as areas such as forests or grasslands. Hello agroecosystems! Hello urban ecosystems! Nor does it constrain us to think of ecosystems as only being large in size, the microbiomes in our guts may also be damaged and need restoring with some yogurt after we've taken an antibiotic. The definition doesn't even confine us to only having native species in an ecosystem.[8] If we value ring-necked pheasants in United States' grasslands as a culturally significant species, we can work to keep them there despite the fact that they are originally an Asian bird. This definition doesn't tell us ecosystems were healthier in the past – although they no doubt were – and we should use some past ecosystem as a baseline for our restoration. This is important because of all the global changes we examined in Chapter 7. We can't go back, we can only go forward. Global environmental change suggests that going forward, many ecosystems will not have historical examples. Future ecosystems will often contain new combinations of species. Healthy ecosystems of tomorrow may be very different from what was present at a given location in the past. Also, what combination of species works will depend on the context of a given site. What works for a **bioswale** lining a city street will be very different than what we want in a local park or nature preserve. The SER's definition tells us only that nature and her ecosystems are sick. They use the three Ds (degraded, damaged, or destroyed). It is our job to help these sick ecosystems recover.

Ecosystem restoration will look different in different places influenced by the local ecology, culture, and history. The basis for any food chain is the producers, and they need clean water and healthy soil/sediments to thrive. Creating these conditions for local plants (or phytoplankton) is always a good first step as is planting plants that were historically in a region (**native species**). Such plants are adapted to the local climate and feed the local animals, so they are key to restoring a system's health. In some cases, restoration will entail stopping harm and remediating past harms. Just like the patient who has smoked for decades, certain landscapes are polluted and need to be cleaned as the first stage in their recovery. On a global scale we know what our goals are, so local restorations should feed into these objectives. As a reminder, these goals are getting the planet back to safe planetary boundary conditions (Figure 7.1), whether that means addressing the climate crisis or saving biodiversity among other global issues (Chapter 7), and continuing the work towards the United Nation's 17 Sustainable Development goals (Table 8.1). Because these are difficult goals to achieve, they need to be approached holistically so progress can be made on many goals simultaneously. Viewing these challenges from an ecological and systems perspective will help us build comprehensive solutions. Groups are already making the necessary linkages across individual goals. For example, the **one health** approach crafts solutions with an understanding of the interdependency of human health with the health of the planet.

Community-Led Conservation

Saving the earth's biodiversity is key to keeping nature resilient. Historically, conserving the earth's plants and animals was often approached by securing land for reserves, parks, or what some have dubbed conservation "fortresses." Lands and waters that had high levels of biodiversity were identified and then preserved and protected for nature. Human use of these areas was typically very restricted to only include certain types of recreation such as hiking. Disturbingly, people were sometimes forcibly removed from their lands in the name of conserving the wilds or wilderness. For example, displacement of Native American tribes is part of the dark history of establishment of the United States' first national parks in the late 19th century including flagship parks such as Yellowstone and Yosemite. Although parks and other protected areas are a critical and necessary piece to any global strategy that aims to stop biodiversity loss, the amount of area they cover is not enough to protect all of earth's species. In addition, parks are not the best conservation approach in many landscapes, particularly landscapes with heavy human demands. If we think of sustainability broadly as working towards all of the UN's sustainable development goals, we have to embrace the idea that community is central to environmental restoration, and this includes not only the biological community but also the human community. Excluding people from preserves can create problems, particularly when exclusion of the local community results from actions by more powerful external groups. Take for example the Chagos Archipelagos in the Indian Ocean, where conservationists lobbied the United Kingdom to create a no-take marine protected area. This area protected marine biodiversity, at least in the short term, but at the cost of marginalizing the Chaggosian community who were not involved in the decision-making (Jeffery, 2014). Ecosystem restoration and biodiversity conservation need to be a social endeavor; any given project will be more likely to succeed if local people are involved at all stages from design to implementation.

The social science behind community resource protection is fascinating. In the best cases, communities will self-organize to protect their local environment including specific species that are of cultural or economic value to that community. In cases where such self-organization occurs, rules for interacting with the natural ecosystem are developed from *within the community* rather than having rules imposed on the community from outside. It is useful to understand when such community self-organization can occur, because self-organized communities are no doubt one of the best approaches to meeting long-term sustainability and conservation goals. However, achieving this outcome is tricky; it requires a very special set of circumstances for communities to self-organize in this way. Nobel prize-winning economist Elinor Ostrom (2009) identified ten factors related to successful community organization, particularly looking at when this led to successful management of natural resources. Briefly, here are the characteristics that make a difference. Size matters. There is an optimal intermediate size for both a managed ecosystem and for the number of resource users. Organization needs to be both worthwhile (not too small an ecosystem, not too few users) but not unwieldy (not too large, too many). In addition to size requirements, a community has to strongly value the resource and perceive its potential scarcity. If the ecosystem dynamics of the system of concern are predictable, community self-organization is more likely because implemented rules will lead to more reliable outcomes. In other words, it helps if restricting use leads to a predictable increase in the resource. In addition, sedentary resources are easier to manage successfully. This makes sense as migration and movement can mean that there is a separation between those who restrict use and those who benefit from the restriction. Respected leaders, shared values, and societal norms all help, as does knowledge of how the system works economically, socially, and ecologically. Finally, success is also more likely if resource users have autonomy to create and enforce their own rules. If you are involved in community-led conservation, or any type of regional scale sustainability project, it pays to reflect on whether these conditions are in place and, if they aren't in place, how they might be cultivated. Working to strengthen these attributes within our local and regional communities can help us create effective change. In some cases, there will be little hope of successful community self-organization to conserve local environments. In these cases, policies would need to be imposed from outside to achieve environmental goals.

Communities are more likely to make sacrifices for and work to preserve natural systems and species they value highly. You protect what you love. As we work for change in our communities, one item to identify is what anthropologists call **cultural keystone species** (Garibaldi and Turner, 2004). We already saw an example of such a species in the last chapter when we looked at the Koala Beach housing development in New South Wales, Australia – which is designed to meet the needs of both people and koalas. In many cases, cultural keystones are intricately woven into a particular culture because these species are the plants or animals used for food, medicine, or to make every day items. Within indigenous groups, which are the strongest examples of sustainable livelihoods, cultural keystones abound. These include species such as wild rice, sweet grass, salmon, caribou, or bison. Part of conservation, then, is rekindling our relationships to other species and creating landscapes that work well for many creatures. This is easier for indigenous groups because, as Robin Kimmerer (2011) writes, "what conservationists call biodiversity, traditional peoples call kin." Despite this familial view of other species in indigenous philosophies, respect and relationship often explicitly involves hunting or harvesting cultural keystone species. In this way, indigenous perspectives toward the earth's fellow species are markedly different from John Muir's "leave only footprints" view that many Western environmentalists are familiar with. Use of a cultural keystone, whether it is cultivation and harvesting or hunting, is a crucial aspect of the relationship, often viewed in these cultures as necessary to cultivating a strong bond. Respectful use strengthens the relationship. This is more than a cultural philosophy. For example, experimental work has shown that sweetgrass reaches higher abundances when it is harvested according to traditional practices compared to when it is left untended. Humans help this species thrive. What I love about examples such as this, is that they show people as positive forces in nature. Environmental science is so weighed down by examples of environmental degradation and loss that we can forget this important truth – our species can and has interacted positively with natural systems. Doing so helps not just wild plants and animals but also us. Robin Kimmerer calls this "reciprocal restoration." We help other species, and they help us. Restoring human-nature relationships turns out to also be incredibly good for both mental and physical health. Restoring the environment restores ourselves. Multiple studies support the positive effects of increasing green spaces in our daily environments. People respond positively to time spent in nature including improved mental health and even decreases in cardiovascular disease and cancer mortality (Box 9.1).

Box 9.1 Touching grass: Nature and human health

Most of us have heard someone say, "I need to go touch grass." It is a colloquialism that means the person needs to get outside and, usually, away from intrusive thoughts and stress. There is more to this saying than meets the eye. An ever-increasing body of research is demonstrating just how good nature is for human health and not just cognitive health. Scholars have done controlled experiments using concrete landscapes as the control. Bratman and others (2015) randomly assigned participants to a 50-minute walk in either an urban (think roads, sidewalks, and traffic lights) or natural environment (e.g., park with grass and trees). They took psychological and cognitive data both before the walk and after so they could determine the effect on mood and cognitive performance. Specifically, they wanted to tease out whether having more nature while walking influenced impacts. This well-designed experiment was able to show that the nature walk had a stronger, positive effect than the urban walk – reducing rumination and anxiety, increasing mood, and increasing memory. Why does being out in nature make us feel better? In part, nature is a multisensory experience. We hear bird song, chattering squirrels, the wind in the trees; we see flowers or perhaps snow dripping from pine needles; we smell all the chemicals that plants and soil emit as they do the work of nature; we feel a slight breeze on our skin, maybe we even touch grass (Franco, Shanahan, and Fuller, 2017). I've left out taste, but if you have an apple in your bag, you can munch it while you stroll along. Work such as Bratman's (2015) highlights the importance of bringing more biodiversity into our cities so this restorative experience is available to all. In addition to providing nature sounds, vegetation also dampens the noise of the city bringing with it a sense of peace.

Although more work has been done on mental health benefits, health impacts are likely to extend beyond our minds. Bikomeye and others (2022) did a systematic review of 31 studies that examined how different types of nature experiences impacted cardiovascular health as well as cancer. The studies were conducted in 11 different countries. The types of activities included in this review ranged from gardening, forest bathing, green exercise, and nature viewing. The authors pooled these together under the umbrella term nature-based interventions (NBIs). This review concluded that such NBIs showed positive effects on both cardiovascular health and effects associated with cancer outcomes. Bottomline – get outside, it's good for you!

Not all of us are embedded in cultures that treat other species respectfully. This means many of us will have to find our way back to these relationships. There is something to be said for picking one species and going all in using your personal set of talents. Clay Bolt personifies this approach: an artist who used his gift for photography to save a small, winged creature: the humble rusty-patched bumble bee. Bolt first worked to produce a short documentary film on this bee's rarity and plight. He and others then petitioned the United States Fish and Wildlife service to give this rare bee legal protection – and won this protection in 2017. Bolt likes to point out in interviews that he is not a biologist. He just learned to see what E.O. Wilson has called "the little things that run the world." Now he works as both a photographer and insect conservation advocate (Waterman, 2011). Although Bolt took this on as his personal career and calling, communities have also taken the approach of identifying a critical species that they then work together to protect. One example of such community conservation is the Green Hairstreak Corridor in San Francisco (Nature in the City). Green hairstreak butterflies are a native species in California and require particular species of native plants to both feed their caterpillars and serve as nectar sources. In San Francisco, the local community has created stepping stones of native plant patches that support the butterfly down a six city-block stretch. Community workdays help to maintain these sites. Backyard nurseries (literally just volunteers growing plants at home) help grow the plant species needed to maintain the habitat patches.

Nature is Everywhere: Conservation in Cities and Farmland

Many natural areas and their ecosystems are now what ecologists call **novel ecosystems**. These are ecosystems that, due to human impacts, differ substantially from those that occurred in an area historically and can't easily be moved back to a system similar to the past – they are resilient in their present, altered state. MacDonald and King (2018) argue that the novel ecosystem concept is a bridging concept that allows cultural and ecological goals to be worked out at a community level. They write that for such ecosystems, "there are no default, ready-made goals for restoration projects. Instead, goals must be defined in a purposeful, self-conscious way, and practitioners are challenged to accomplish this in a way that is equitable and just." Novel ecosystems make us ask questions and share ideas. What can be conserved or restored at this place? What attributes does our community value? What is our vision for a future healthy ecosystem? Is there a particular species whose conservation we could aid? If nature is everywhere, conservation can also occur everywhere although it will take on many different guises.

One size does not fit all. Conservation and restoration will be place and people specific. Green hairstreak butterflies can be the focus in San Francisco and koala bears in New South Wales. Different approaches are and will be needed as social and ecological systems shift. Community involvement and commitment is a must. With just a little help, nature can thrive in unexpected places: on roofs, in farm fields, in cities. Rivers can flow through cities where once only paved roads ran. Birds of prey can swoop down from the urban sky and catch an unsuspecting pigeon. The wonderful news is that when we help other plants and animals thrive in our landscapes, we get lots of additional environmental benefits. We find that air and water pollution decline as does the amount of storm water runoff, thus decreasing flooding risk. In the case of living roofs, we get cozy insulated buildings that help decrease our electric bill. Nature just needs a little bit of soil or a little bit of water to take hold. Michael Rosenzweig (2003) wrote "Win-Win ecology," where he advocated for a "third strand" of conservation where we increase biodiversity everywhere we can. Rosenzweig specifically is talking about human dominated landscapes not nature reserves or other locations where preserving or restoring biodiversity is already pursued. He called this approach **reconciliation ecology** and opens his book with these comforting sentences: "There is still time. There is good reason to believe that civilization need not destroy most of the Earth's nonhuman species. The trick is to learn how to share our spaces with other species" (opening sentences of Preamble, p. ix).

In the city, a reconciliation approach would include replacing lawns with gardens that are rich in native plants and/or grass types that don't need mowing. Entomologist Doug Tallamy has multiple books advocating and describing this approach (2020). Reconciliation would also include establishing living roofs and, in some cases even living walls. It would mean restoring natural creeks and waterways. Sometimes such creek restoration means "daylighting" these waterways – bringing creeks back to the surface instead of banishing water to the sewer systems. Ideally, these new urban ecosystems will support as many creatures as possible. This is the power of using the plants that are native to any given region. Native plants are adapted to the local climate and soils and are the food source for the local fauna. Insects make up the vast majority of animal biodiversity, and they need much smaller areas to thrive in contrast to the bigger animals we tend to associate with "wildlife" conservation. Sometimes butterflies are just passing through and need a little sip of nectar. Your little patch of hope can be a pit stop for them on their long journey. Amazingly, living roofs can support *so many species*. Gyongyver Kadas (2006) found hundreds of species of spiders, beetles, ants and bees living in London's living roofs. Incredibly, some of these were even rare insects. She also found lots of snails which along with insects and spiders make great bird food. This work means that adding just 10 cm of soil and a suite of hardy plants to any rooftop is a boon for local animals. Roofs are everywhere which means the potential for living roofs as habitat is tremendous. Another ubiquitous urban, suburban, and rural landscape feature is, of course, the lawn.

A lot of land is locked up in lawns. They take up an estimated 2% of land in the United States, over 40 million acres. Although lawns can support some species, if they are mowed frequently and managed with fertilizers and herbicides, they don't do much for local biodiversity. They also contribute to water pollution and **eutrophication**. Mowing releases greenhouse gases and lawn equipment is notorious for emitting air pollutants. There is a reason environmentalists and conservationists have raised awareness about lawns – they have enormous potential given their size and current use. They represent an opportunity to shift a substantial part of many landscapes away from a harmful land use and to a restorative approach that not only helps curb biodiversity loss but also decreases nitrogen/phosphorous pollution. They are also frequently *right there* in front of our houses, blanketing our college campuses, around our business sites. What's stopping us from killing them and bringing back some wild? Societal norms are stopping us, which means we just have to change our minds about what we want the landscapes to look like where we live and work. Replacing lawns with native plantings helps address at least three of the environmental crises that Richardson and others (2023) found where in the danger zone (Figure 7.1).

Transforming lawns is easier than ever with so many groups eager to provide guidance such as Chicago's Rethinking Lawns Project (n.d.).[9] Local native plant societies are strong advocates for changing lawns into more biodiverse plantings. In the eastern United States, there are multiple Wild Ones chapters that do this work. What might a native planting look like? It depends on what you and your local community want in any given space. Are there aesthetic goals you want to achieve? Cultural goals and uses for the area? How tall do you want the mature plants to be in your local planting? Whether it is in your own yard or at a community park, school, or local business – you can design plantings based on your needs and desires. Maybe you only want short plants because you want people to be able to see past the planting to admire a local piece of artwork. Maybe you want certain colors of flowers. Maybe there are some local butterflies or beetles you want to make sure you have the host and nectar plant sources for. What might serve as a **cultural keystone species**? The fun thing about ecosystem design in an area that has been a lawn or a roof is you don't have to try to return it to some prior system. You can decide what you want and why. Almost anything you do will improve things! Philadelphia biodiversity advocate Sarah McAnulty[10] created colorful stickers that say: "Literally just plant something. Pick something native and plant it." There are a few best practices – stick with native plants, it's good to have a variety of plants that flower across the growing season so your pit stop always has fuel, put up landscape edging and a sign so that people know what you're up to. These aren't weeds, I'm saving the world! As Dr. McAnulty's sticker suggests the one big rule is to focus on native plants for your region. The reason for this is of course the evolutionary history between an area's native plants and the local animals you want to support. Native plants provide both the food and habitat for native animals. Even this rule, however, can be bent. Your landscape doesn't have to be 100% native plants, 75% is fine. Go ahead plant some tulips or rose bushes. By all means, put in that vegetable garden as well and a food forest (orchard). The topic of food brings us to growing food for ourselves and others. I live in Central Illinois in what my students have called "the corn desert." Are corn and conservation compatible? Maybe more than you think.

Grassland biomes are at very high risk of conversion worldwide. They are considered an endangered biome because of high loss and low levels of protection (Hoekstra et al., 2005). Across the planet we have plowed up grasslands, transforming them to agricultural fields. This is very true for my home town which sits nestled in a sea of corn and soybean fields. What was once tallgrass prairie is now row crop agriculture. Corn is after all a highly modified grass, so it isn't too surprising that it grows well where native grasslands once thrived. Industrial agriculture has a suite of negative environmental impacts and is a major reason we are past the planetary boundary for the amount of reactive nitrogen and phosphorus entering our rivers and oceans. Reconciliation ecology has a solution for this. Prairie strips. Prairie strips are just what they sound like, small strips of native grasses and herbaceous plants that are planted along the elevational contours of a farm field. These strips don't dominate the field. In the areas of Iowa where they've been added they take up 10% of the field, with the remaining 90% planted as typical row crop agriculture. But what a difference that 10% makes! Lisa Schulte and others (2017) worked with Iowa farmers to determine how adding 10% of prairie into their fields would affect the environmental impact of their fields. Here's what they found: insects and birds, including birds of conservation concern and pollinators, increased two- to three-fold. There's more: stormwater runoff from fields decreased 37% which resulted in more soil and phosphorus being left on the fields, 20-fold and four-fold increases respectively (Schulte et al., 2017). Truly phenomenal! I am a prairie strips cheerleader and am happy to dress up as a pollinator and sing their praises anywhere at any time. Farmers, of course, would want to know how adding these strips affects crop yields. Again, good news: no change in yield once the decrease in 10% is accounted for. This practice makes so much sense, and we need to find the environmental policy tools to put it in place everywhere farming takes place. People want the environmental benefits that prairie strips provide. Schulte looked at this, as well, in her study – surveying people including farmers on their opinions. People want this so it is up to us to figure out how to incentivize getting it done. It's a systems issue. How do we redesign agricultural and conservation policies so that such work gets done?

Prairie strips make sense for Midwestern corn fields, but agricultural conservation approaches will of course vary by region and crop. Some crops can actually be grown interspersed with other plants. One example of this is tropical shade-grown coffee. Historically, coffee was grown much like corn, in full sun as a monoculture. However, coffee trees grow well in the shade and growing coffee beneath native shade trees has tremendous ecological benefits. This approach supports birds, ants, and other wildlife including pollinators. In turn, this increased wildlife naturally controls pests in the plantations meaning less need for synthetic pesticides. In addition to all that, increased pollination, carbon sequestration, improved water quality, and improved bean quality have been documented in studies. Perhaps due to higher rates of pollination, coffee yields are comparable between shaded, biodiverse plantings, and sun-grown coffee plantations (Wright et al., 2024). Same yield, better for the planet, sounds good to me. As always, systemic changes are important. In the case of coffee, small labels on coffee, that indicate it is shade-grown (ecolabeling), help consumers identify and support better farming practices.

The more you start to think about everywhere as ecosystems and potential habitat, the more potential restoration spots you'll see as you make your way through the world. Roadsides are another fantastic opportunity for increasing habitat, in part because larger roads already have natural right-of-ways, ripe for more biodiverse plantings. As a bonus, roads traverse the land, naturally connecting areas to one another. Having the United States' highways bloom with wildflowers is not a new idea but implementing a change from a system designed to have only lawn grass along roadways has proved challenging. In the United States there is legislation to promote this change, the 1965 Highway Beautification Act. First lady Lady-bird Johnson loved nature and hated the billboards and trash that she saw along the highways. Because of her love of flowers and her love of President Lyndon Johnson, she was able to play a role in changing the United States' highways. The goal of the Highway Beautification Act was and is restoring natural beauty to US highways. Over time, this idea changed to embrace a conservation goal as awareness of declines of beloved species such as the Monarch butterfly, which summers in the United States, grew. In 2015 the Federal Pollinator Task Force recommended planting pollinator plants along North-South highways which would connect habitats from Texas to Minnesota. Despite these initiatives and policies many US highway rights-of-way are still primarily grass. Some states, such as Iowa and Texas, do a better job than others at roadside restoration but there is still much room for improvement. There is no reason for highways everywhere to not have biodiverse natural strips along them. What a treat it would be to learn the native flora of a place as we drive, bike, or wander through the land.

To the extent possible, conservation objectives can and should be coupled to other sustainability goals. Some fields filled with solar panels can and are doing good work beyond providing a clean renewable energy source. The land beneath and between solar panels can be planted in a diverse mix of native wildflowers and thereby support local insect, including pollinator, populations, which of course support the local bird populations, and so on (Blaydes et al., 2021). Wires move electricity generated in solar farms and wind farms across the land and above-ground electrical lines also have rights-of-way beneath their towers. Here again are areas with ecosystem potential. Once again native plantings are the key. Keep your eyes peeled as you walk the land or kayak the waters. Is there soil? Could it grow something more than grass or weeds? I've restrained myself from a deep discussion of **guerilla gardening** so as not to implicate myself, but it turns out that breaking the law for flowers is a thing (Reynolds, 2014). Much like street art, the idea here is to enhance the plants growing in neglected spots through illicit seeding. Ecological graffiti anyone?

Snapshots of Success

In my hometown of Springfield, Illinois, there is a small 100-acre property named Lincoln Memorial Garden (LMG). It turns out, it is not a garden at all: it is a forest and prairie with hiking trails throughout, a play area for children, and a nature center. Recently, large wooden trolls have been constructed to add art and whimsy to LMG. It has a fascinating history, spearheaded by a local firebrand named Harriet Knudson who insisted that landscape architect Jen Jensen design a living memorial to US President Abraham Lincoln here in Springfield. (For those of you not from Springfield, Lincoln practiced law and served in the state legislature in Springfield from 1837 until his election to the presidency in 1860.) Knudson was like a dog with a tennis shoe in her persistence. She got the city to lease city property to the garden club for $1, she called for white oak acorns to be shipped to her address from state garden clubs, gathering them from 28 different states. In 1936, boy and girl scouts planted acorns throughout the property which at that time was primarily grass with a few scattered trees. Today LMG has a small oak and maple forest along with prairie plantings that are teeming with life (Lincoln Memorial Garden). It is a favorite spot for hiking, birdwatching, and outdoor events. They run a small summer camp for kids. A Monarch butterfly that was tagged during an educational event in 2016 was found 1800 miles south in El Rosario, Mexico. This property is a small site, in a small town. There could be more living memorials such as this throughout the world. What local historical figure might your town create a living memorial for?

Half way around the planet from my home is the tiny island country of Singapore. This country has one of the highest population densities in the world. It sits just a smidge above the equator and therefore has a warm, wet climate year-round. Both on land and in the sea, this small country is doing so much right. They call it **biophilic design** (Er, 2021). In the sea, they are working diligently to restore their coral reefs, tracking where the baby corals float to so the reserve network will work as it should. On land, parks serve as core zones for protecting biodiversity with corridors between them that are created so as to be the paths of least resistance for fauna. But they have taken sustainability beyond their reserves and natural parks embracing a "City in Nature" vision. In 2020, they pledged to plant one million trees by 2030. Across urban areas native plantings are designed to "emulate natural forests," to enhance health and well-being they are designing "therapeutic" natural landscapes and "nature play gardens" for children. They are working to restore their rivers and lakes by transforming concrete canals back to vegetated waterways with floodplains. They are a world leader in "vertical greening" with plants dripping from their buildings and springing from their patios and balconies. New York City may be proud of the native plantings along the highline but clearly it still has a long way to go if it wants to be a "City in Nature."

I am an animal and plant lover at heart. As a child I would race around my backyard in California attempting to catch lizards and would be surprised when I ended up with just a tail in my hands for the effort. Perhaps because of this, the success stories that speak loudest to me are those that center around individual species. To reiterate what I've said already – I endorse falling in love with a species and then helping to make the world do a better job of meeting that species' needs. We saw how photographer Clay Bolt picked the rusty-patched bumble bee, and how the San Francisco community picked the Green Hairstreak. In Mexico City, a different story is unfolding with passionate conservationists working to shift an urban ecosystem into one with restored wetlands that support the Mexican axolotl. It's easy to fall in love with an axolotl – salamanders that retain their juvenile characteristics of external gills (Figure 9.3). So, it was upsetting to many when, despite the Mexican axolotl's abundance in captivity, it seemed as if this rare amphibian was well on its way to extinction in the wild as wetlands disappeared due to farming and housing developments (Contreras et al., 2009). Luis Zambrano and others, however, approached conservation of this species from a systemic perspective – knowing that long-term preservation of the axolotl around a densely populated city, Mexico City, would require a resilient social-ecological system (Zambrano et al., 2020). Xochimilco refers to a canaled farming system which is found south of Mexico City. These canals have been severely degraded over time. One consequence of this degradation was disappearance of Mexican axolotls from associated wetlands. The wetlands of Xochimilco, called Chinampas, were historically

Figure 9.3 Adorable Mexican axolotl
Note: This endangered salamander with its distinctive external gills is found in the wild only in Mexico near Mexico City.
Source: Amandasofiarana (2014).

farmed but as traditional farming declined the degradation of these wetlands increased. The Chinampa-Refuge project uses the salamander as a focal species and combines salamander conservation with renewal of historic farming practices. The hope is that this will create a system where Chinampas, Chinamperos (traditional farmers), salamanders, and water quality all thrive. Captive populations of axolotls are bred to increase their numbers. Chinamperos create salamander refuges (watery ditches) on their lands for reintroduction of the species (Smith, 2023). Salamander reintroductions are still in their early stages but show initial signs of success with individual axolotls surviving and growing in these traditional farm refuges.

Art, Anxiety, and Activism

I've discussed throughout this book how new terms have arisen as the scale of environmental challenges has revealed itself: **Anthropocene, biodiversity, sustainability, wicked problems** to name a few. Recently, psychologists have coined two additional terms to add to this growing list: eco-anxiety and **climate anxiety**. These phrases are self-explanatory in nature, describing the mental health issues that are arising from heightened knowledge of environmental crises. The younger generation feels the burden put on it from the system they find themselves in and often feels powerless to affect change. I see this in the eyes of my students when I cover the science. They've heard the science before but they haven't seen society shift significantly in response. It turns out that eco-anxiety is thought to be a type of anxiety that psychologists call a practical anxiety, where the concern makes individuals seek out information, alter behavior, and act (Kurth and Pihkala, 2022). In the art world, one result is a phenomenon called eco-art – art which aims to bring attention to ecological issues and foster action. Sometimes this work involves creating art from what would otherwise be waste. For example, Canadian artist Gerald Beaulieu constructed large black crow sculptures made out of rubber tires, which he placed sprawled across the ground as if dead (Ross, 2023). These sculptures made some Canadians uncomfortable but the goal of the art was not to provide comfort but to raise awareness and stimulate reflection. I imagine you weren't expecting to hear about dead crow sculptures in a chapter on hope; but seeing such art does bring me hope. These works challenge people to consider our interaction with the natural world.[11] Don't look away. Hopefully they also motivate people to act. Action helps both decrease anxiety and foster hope – whether the action is starting a community garden or creating a work of art (Figure 9.4). Such actions plant the seeds of hope and each of us has our own unique set of skills and joys to bring to the task (Johnson, 2024). (I have sung the praises of Dr. Johnson's book multiple times in this book but to add just one more tantalizing fact: it contains an anti-apocalypse playlist.) Redesigning unsustainable systems is hard work, some levers work better than others

Figure 9.4 The forest is watching you
Source: Tirapolsky (2012)

at affecting change (Meadows, 2008). We will undoubtedly need good music for this work, along with all other forms of art. We will need this work to both comfort us and push us out of our comfort zone.

Conclusion

On a tree planting day, standing on a median strip on a busy Springfield street, I first met Anne Logue, a thin, older woman with short, grey hair, she walked decisively toward our group pick-axe in hand. In some places, getting to the soil means removing asphalt and Anne came prepared. This memory always makes me smile and it is one that gives me hope. *Don't count the old ladies with pick-axes out!* How long will it take to achieve the UN's sustainability goals? Will we get our planet back to within the safe zones for all the planetary boundaries we've crossed? If so, when? How soon? I don't know the answers to these questions. Some days I'm more hopeful than others that it will happen sooner, rather than later or not at all. Some days I'm pulled into despair when it seems as if greed will consume the world leaving nothing for those of us who wanted just a simple, healthy life for ourselves and our families. What I can say is hope is always a better choice than despair so I keep forcing myself to choose hope. "Hope is a verb" is not a bad mantra. Until I no longer can, I will roll up my sleeves, grab my pick-axe and/or laptop, and keep trying. My best days are the days I can work with joy and celebration, with gratitude towards the earth that feeds and nourishes us. Sometimes I am able to laugh and hope that the earth will forgive us and accept our work as a long overdue apology. The most delightful times are when I can feel community building around me as we place a tree's roots in the ground.

Notes

1 Dr. Ayana Johnson has done marvelous work, focusing on solutions to the climate crisis and working on imagining positive futures. I highly recommend reading her book which consists of interviews with people involved in all sorts of endeavors, but who share the common goal of addressing the climate crisis (Johnson, 2024).

2 We saw in Chapter 8 how many indigenous people view human-nature relationships as one of reciprocity, earth gives us gifts and we have a responsibility to reciprocate that gift through care and stewardship.

3 Fascinatingly, this is not the wording Aldo Leopold used in his famous book *The Sand County Almanac* published in 1949 (Leopold, 1970). This quote which includes people explicitly as part of the land community comes from an earlier article he wrote for the Garden Club of America (Leopold, 1947).

4 Nature is frequently referred to as Mother Nature and therefore given a feminine pronoun. A good friend of mine, Dr. Richard Pamenter, once pondered whether this was actually good for nature conservation given sexism. I've used feminine or neutral descriptions throughout this book but you might think about how you might view nature differently if it was Father Nature who couldn't care less.

5 Although some marine mammals such as the Vaquita porpoise are severely at risk of extinction, other species have recovered to population sizes ranging from 10,000 for blue whales to 80,000 for humpback whales (Mehta, n.d.).

6 House finch bill changes are by no means the only example of evolution in urban environments which is now very well documented. To find out more about this fascinating phenomenon read Johnson and Munshi-South (2017).

7 For more information on this United Nations declaration and associated projects see UN Environment Programme in references (n.d.).

8 The definition used here is from SER's (2002) primer which does not mention native or nonnative status. In contrast, a more recent definition on the SER's website has been updated to read "recovery of a native ecosystem" which does decrease the scope of what would be considered ecosystem restoration by this academic society.

9 Researchers in Chicago are studying how lawns compare to various types of native plantings in their ecological and social outcomes. Find out more about the work of Dr. Becky Barack and others at the project's website (Rethinking Lawns Project).

10 Dr. McAnulty also drives around Philadelphia in a car named the Squid mobile which is decorated with paintings of squids and has a phone number to text to receive squid facts. How do we get more people like this in the world?

11 One example of art that addresses the ecological crisis is the play *Hurricane Diane* by playwright Madeleine George (2019). In this play the Greek demigod Dionysus returns in human form to recruit followers to help restore nature. The god/goddess selects four housewives in a New York suburb to charm and seduce. George's play hilariously captures the psychological impulse to deny the realities of global environmental change particularly if this reality means we must do something truly dramatic like … ripping out our lawns.

References

Amandasofiarana. (2014). Axolotl Abystoma mexicanum. Wikimedia Commons. https://commons.wikimedia.org/wiki/File:Axolotl_ambystoma_mexicanum_anfibio_ASAG.jpg.

Badyaev, A. V., Young, R. L., Oh, K. P., and Addison, C. (2008). Evolution on a local scale: developmental, functional, and genetic bases of divergence in bill form and associated changes in song structure between adjacent habitats. *Evolution*, 62(8), 1951–1964. doi:10.1111/j.1558-5646.2008.00428.x.

Bikomeye, J. C., Balza, J. S., Kwarteng, J. L., Beyer, A. M., and Beyer, K. (2022). The impact of greenspace or nature-based interventions on cardiovascular health or cancer-related outcomes: A systematic review of experimental studies. *PLOS One*, 17(11), e0276517. doi:10.1371/journal.pone.0276517.

Blaydes, H., Potts, S. G., Whyatt, J. D., and Armstrong, A. (2021). Opportunities to enhance pollinator biodiversity in solar parks. *Renewable and Sustainable Energy Reviews*, 145, 111065. doi:10.1016/j.rser.2021.111065.

Bratman, G. N., Daily, G. C., Levy, B. J., and Gross, J. J. (2015). The benefits of nature experience: Improved affect and cognition. *Landscape and urban planning*, 138, 41–50. doi:10.1016/j.landurbplan.2015.02.005.

Contreras, V., Martínez-Meyer, E., Valiente, E., and Zambrano, L. (2009). Recent decline and potential distribution in the last remnant area of the microendemic Mexican axolotl (*Ambystoma mexicanum*). *Biological Conservation*, 142(12), 2881–2885. doi:10.1016/j.biocon.2009.07.008.

Er, K. (2021). Transforming Singapore into a city in nature. *Urban Solutions*, 19, 68–77.

Franco, L. S., Shanahan, D. F., and Fuller, R. A. (2017). A review of the benefits of nature experiences: More than meets the eye. *International journal of environmental research and public health*, 14(8), 864. doi:10.3390/ijerph14080864.

Garibaldi, A. and Turner, N. (2004). Cultural keystone species: Implications for ecological conservation and restoration. *Ecology and Society*, 9(3), 1. https://www.jstor.org/stable/26267680.

George, M. (2019). *Hurricane Diane*. Acting edition. Concord theatricals. Samuel French.

Hoekstra, J. M., Boucher, T. M., Ricketts, T. H., and Roberts, C. (2005). Confronting a biome crisis: Global disparities of habitat loss and protection. *Ecology Letters*, 8(1), 23–29. doi:10.1111/j.1461-0248.2004.00686.x.

Howells, E. J., Abrego, D., Liew, Y. J., Burt, J. A., Meyer, E., and Aranda, M. (2021). Enhancing the heat tolerance of reef-building corals to future warming. *Science Advances*, 7(34), eabg6070. doi:10.1126/sciadv.abg6070.

Jeffery, L. R. (2014). Ecological restoration in a cultural landscape: conservationist and Chagossian approaches to controlling the "coconut chaos" on the Chagos Archipelago. *Human Ecology*, 42(6), 999–1006. doi:10.1007/s10745-014-9696-y.

Johnson, A. E. (Ed.). (2024). *What if we get it right? Visions of climate futures*. Penguin Random House.

Johnson, M. T. and Munshi-South, J. (2017). Evolution of life in urban environments. *Science*, 358(6363), eaam8327. doi:10.1126/science.aam8327.

Jones, H. P. and Schmitz O.J. (2009). Rapid recovery of damaged ecosystems. *PLOS One*, 4(5): e5653. doi:10.1371/journal.pone.0005653.

Kadas, G. (2006). Rare invertebrates colonizing green roofs in London. *Urban Habitats*, 4(1), 66–86.

Kimmerer, R. (2011). Restoration and reciprocity: the contributions of traditional ecological knowledge. In: D. Egan, E. E. Hjerpe, and J. Abrams (Eds.) *Human dimensions of ecological restoration: Integrating science, nature, and culture* (pp. 257–276). Island Press. doi:10.5822/978-1-61091-039-2_18.

Kurth, C. and Pihkala, P. (2022). Eco-anxiety: What it is and why it matters. *Frontiers in Psychology*, 13: 981814. doi:10.3389/fpsyg.2022.981814.

Leopold, A. (1947). The ecological conscience. *The Bulletin of the Garden Club of America*, 45–53.

Leopold, A. (1970). *A Sand County Almanac*. 1949. Ballantine.

Lincoln Memorial Garden. (n.d.). History. https://www.lincolnmemorialgarden.org/about-us/history/.

Macdonald, E. and King, E. G. (2018). Novel ecosystems: A bridging concept for the consilience of cultural landscape conservation and ecological restoration. *Landscape and Urban Planning*, 177, 148–159. doi:10.1016/j.landurbplan.2018.04.015.

Meadows, D. H. (2008). *Thinking in systems: A primer*. Sustainability Institute.

Mehta, R. (n.d.). Global whale population status: Current insights. Physixis. https://physixis.com/articles/global-whale-population-status/.

Orr, D. W. (2004). Hope in hard times. *Conservation Biology*, 18(2), 295–298. http://www.jstor.org/stable/3589204.

Ostrom, E. (2009). A general framework for analyzing sustainability of social-ecological systems. *Science*, 325(5939), 419–422. doi:10.1126/science.1172133.

Ramirez, A. (2025, May 25). Mexico saves coral reefs with 3D printers – Marine life is back. Engineerine. https://engineerine.com/3d-printed-coral-reefs/.

Rethinking Lawns Project. (n.d.). Rethinking lawns. https://rethinkinglawns.com/home/about-us/.

Reynolds, R. (2014). *On guerrilla gardening: A handbook for gardening without boundaries*. Bloomsbury Publishing.

Rosenzweig, M. L. (2003). *Win-win ecology: how the earth's species can survive in the midst of human enterprise*. Oxford University Press.

Ross, S. (2023, July 16). When the crow doesn't fly: PEI artists accept that not everybody will like their work. CBC. https://www.cbc.ca/news/canada/prince-edward-island/pei-art-reaction-beaulieu-1.6902365.

Schulte, L. A., Niemi, J., Helmers, M. J., Liebman, M., Arbuckle, J. G., James, D. E. ... and Witte, C. (2017). Prairie strips improve biodiversity and the delivery of multiple ecosystem services from corn–soybean croplands. *Proceedings of the National Academy of Sciences*, 114(42), 11247–11252. doi:10.1073/pnas.1620229114.

Smith, J. E. (2023, Dec. 5). What it takes to save the axolotl. *New York Times*. https://www.nytimes.com/2023/12/05/science/mexico-axolotl-biology.html.

Society for Ecological Restoration Science and Policy WorkingGroup. (2002). The SER primer on ecological restoration. *Society for Ecological Restoration*. www.ser.org.

Tallamy, D. W. (2020). *Nature's best hope: A new approach to conservation that starts in your yard*. Timber Press.

Tirapolsky. (2012). Street art in the forest. Wikimedia commons. https://upload.wikimedia.org/wikipedia/commons/2/28/Street_art_in_the_forest.jpg.

UN Environment Programme. (n.d.). Preventing, halting, and reversing loss of nature. https://www.decadeonrestoration.org/.

US National Archives and Records Administration. (1910). Photograph of hardwood logs at mill in Grayling, Michigan. Wikimedia Commons. https://commons.wikimedia.org/wiki/File:Photograph_of_Hardwood_Logs_at_Mill_in_Grayling,_Michigan_-_NARA_-_2129585.tif.

UVI. (2022). Coral restoration. Wikimedia Commons. https://commons.wikimedia.org/wiki/File:Coral_Restoration.jpg.

Waterman, J. (2011). Clay Bolt's quest to save a species with images. https://www.bhphotovideo.com/explora/photography/features/clay-bolt's-quest-to-save-a-species-with-images.

Westervelt, K. (2011). From adversity to diversity: The Cape Florida project. In: D. Egan, E. E. Hjerpe, and J. Abrams (Eds.) *Human dimensions of ecological restoration: Integrating science, nature, and culture* (pp. 39–50). Island Press.

Wright, D. R., Gordon, A., Bennett, R. E., Selinske, M. J., Lentini, P. E., Garrard, G. E. ... and Bekessy, S. A. (2024). Biodiverse coffee plantations provide co-benefits without compromising yield. *Journal of Sustainable Agriculture and Environment*, 3 (3). doi:10.1002/sae2.70005.

Zambrano, L., Rivas, M. I., Uriel-Sumano, C., Rojas-Villaseñor, R., Rubio, M., Mena, H. ... and Tovar-Garza, A. (2020). Adapting wetland restoration practices in urban areas: Perspectives from Xochimilco in Mexico City. *Ecological Restoration*, 38(2), 114–123. doi:10.3368/er.38.2.114.

Glossary

3Es	stands for the combined objectives of sustainable development: environment, equity, and economy
Abiotic environment	non-living components of the world around us (e.g. soil, air, water)
Adaptation	a characteristic that increases survival or reproduction in a given environment
Adaptive evolution	increase in genes that code for adaptations across generations due to higher survival and reproduction of individuals with those genes
Adaptive management	an iterative management approach where new approaches are tested and are changed in response to increased understanding of the ecosystems being managed
Anthropocene	a new geologic epic characterized by human dominance and alteration of the planet
Assisted Evolution	aiding the adaptive evolution of a species by developing and introducing genotypes that do well in a human altered environment
Biodiversity	the diversity of life across all biological scales including genetic, species, and ecosystem diversity
Biomagnification	when natural processes increase the concentration of fat-soluble chemicals as they make their way up the food chain creating greater risk of toxic effects for top predators (e.g. DDT and eagles)
Biomes	large terrestrial regions that are characterized by distinct climate and associated vegetation type (e.g. temperate grassland, tropical forest)
Biophilic design	using nature as a guide for urban planning and design including enhancement of biodiversity
Bioplastics	pliable materials that are often biodegradable some of which have natural materials as their source
Bioswale	natural plantings that use vegetation to absorb rainwater and storm runoff

DOI: 10.4324/9781003725404-10

Brandolini's Principle	it takes considerably more energy and time to effectively debunk misinformation than it takes to create and spread misinformation
Cap-and-trade	policy that sets a cap on a certain activity and allows selling of credits among companies for rights to perform such activity
Carbon footprint	amount of atmospheric carbon a person or country creates in a given period of time
Carbon sinks	ecosystems that have a net absorption of atmospheric carbon
Carbon tunnel vision	tendency to focus on the climate crisis rather than other planetary-level environmental changes that put the planet at risk
Cause-effect relationship	when one factor is responsible for a given outcome (e.g. lower temperature causes ice to freeze)
Chemical energy	energy contained within molecules such as sugars or protein
Circular economy	an economy that avoids waste by creating products that reuse materials from previous products
Click-restraint	stopping oneself from sharing information before you have fact checked its validity
Climate anxiety	mental health effects that arise from concern about the climate crisis
Community	1. an ecological term referring to a group containing many different species that are found together and interact (e.g. grassland community of plants and animals) 2. a group of people who live in a given location and therefore have a shared interest in the location's environmental health
Competitors	an ecological term referring to two species that negatively influence one another due to use of a shared resource
Confirmation bias	people are more likely to believe information that is consistent with their existing worldviews
Consumers	an ecological term referring to organisms that get their energy through consuming producers, other consumers, or both.
Continued influence effect	continued influence of early misinformation on a person's beliefs even after they have acknowledged its falsehood
Controls	experiments can have both negative and positive controls. A negative control is run with the same conditions but omitting the treatment being tested. A positive control uses an established

	treatment that already is known to produce a given effect.
Coral bleaching	when coral expels the symbiotic algal associate that gives it is color (see coral reefs for more on coral biology)
Coral reefs	ecosystems that are dominated by coral which is formed by a symbiosis between colony forming invertebrate animals (form coral's structure) and algal associates (give coral color).
Cultural keystone species	species that play a significant role in a given culture
Cultural services	a subgroup of ecosystem services, specifically referring to the psychological, recreational, spiritual, and educational benefits people get from nature
Decomposers	organisms, typically bacteria and fungi, that break down complex biomolecules into simple nutrients
Detritivores	organisms that eat dead organisms and therefore break the material down into smaller and smaller pieces (e.g. scavengers, soil invertebrates)
Disinformation	false information spread with the intent to misinform others
Dominant species	species that is most prevalent in an ecosystem based on numbers and size
Doughnut economics	approach to economics developed by Kate Raworth that sets a minimum level of economic activity based on human needs and a maximum based on ecological limits
Eco	Latin word root meaning home
Ecology	science that studies nature, specifically the interactions of organisms with each other and with their abiotic environment
Ecological disturbance	an abiotic event that kills many of the organisms in an ecosystem in a short period of time and therefore shifts the environment significantly (e.g. wildfire, flooding)
Ecological economics	a subfield of economics that incorporates ecological costs and constraints
Ecological footprint	area of land and water needed to provide goods and services and process waste
Ecological regime shifts	when an ecosystem undergoes a significant change in a short period of time from one ecological state to a second state
Economic externality	an external cost or benefit that arises from production or consumption of a good or service

Ecosystem	a living system in which organisms interact with one another and with the non-living (abiotic) environment
Ecosystem diversity	diversity of life at the ecosystem scale encompassing the variety of aquatic and terrestrial systems
Ecosystem engineer	a species that dramatically alters the non-living aspects of an ecosystem
Ecosystem restoration	returning an ecosystem that has been degrade to a healthier condition
Ecosystem services	all the goods and services ecosystems provide to people
Ectotherm	organisms that cannot regulate their body temperature
Endotherm	organisms that can regulate their body temperature
Energy	capacity for doing work, comes in many forms both compact and dispersed
Energy efficient	uses energy in a way that extracts the maximum amount of work from an energy conversion
Entropy	measure of disorder or randomness
Eutrophication	excessive growth of algae (phytoplankton) in an aquatic ecosystem due to large inputs of either reactive nitrogen or phosphorous
Evolution	change in genetic make-up across generations leading to formation of different taxonomic groups as gene pools separate and diverge
Extinction vortex	when small populations continue to decline in numbers due to a variety of interconnected factors including inbreeding depression, natural variation in sex ratio, Allee effects
Feedback	a process in a system that serves to either accelerate or decelerate its rate
Food web	a diagram of all the feeding relationships among species in a community
Gene	a string of nucleotide code (DNA or RNA) that encodes information such as the sequence of amino acids that form a protein
Generalists	organisms that can survive and reproduce across a broad range of conditions
Genetic diversity	diversity of life at the genetic scale including all the genetic variation found within different species
Geoengineering	applying our understanding of earth's processes for large-scale modification of the earth specifically earth's climate

Global change	environmental shifts that are global in nature including climate change, invasive species, pollution, land use change, and others
Glocal	term used to emphasize the impact of global scale markets and environments on local communities (global plus local)
Greenhouse gases	atmospheric gases that absorb and reradiate heat
Guerilla gardening	adding seeds or plants to neglected areas typically without prior permission
Habitat	areas that meet a species niche requirements
Heuristics	mental short cuts used in decision making
Illusory truth effect	repetition of a falsehood creates familiarity which increases people's acceptance of the falsehood as true
Impervious surfaces	areas that are covered with materials that don't absorb water (e.g. asphalt or concrete)
Indirect effect	an ecological term referring to when a species influences another species through its effects on a third species
Hidden water	water use that is masked because it isn't directly visible to the consumer
Hypothesis	an explanation of how the natural world works
Inbreeding depression	decreases in health and survival of offspring resulting from mating of closely related individuals
Insecticide-resistance	trait that allows insects to survive and reproduce despite presence of an insecticide
Interaction web	a diagram of all the interactions among species in an ecological community including feeding relationships, competitive relationships, mutualistic relationships, habitat relationships and more
Invasive species	pest species that cause damage to ecosystems and affect human well-being (often non-native)
Keystone species	species that have a strong impact on the dynamics of an ecosystem despite their low abundance
Land ethic	ethic proposed by Aldo Leopold that advocates for preserving the integrity of biological communities
Land use change	shifts in an area's use often referring to shifts from natural ecosystems to agricultural, urban, or other human associated uses
Lateral reading	reading information from a variety of sources as you work to determine the validity of a claim
Lazarus effect	when a group of individuals are found of a species previously thought to be extinct

Linear economy	economy that has waste as the end product with goods moving from manufacturing to consumer use to waste
Macroevolution	evolutionary change that produces broad taxonomic differences over time
Market failure	economic term which refers to when a free market creates a loss in social well-being
Mathematical models	bundles of equations that use quantitative data to predict values for certain variables
Mathiness	when a mathematical formula format is used to convey ideas that aren't quantitative thus giving a false impression of rigor
Microbes	a diverse group of very small organisms that are invisible to the naked eye includes various bacteria, fungi, algae
Microevolution	evolutionary change within a species
Microplastics	plastic pieces less than 5 mm long
Misinformation	incorrect information on a topic
Modern synthesis	coupling of Darwin's evolutionary theory with current understanding of genetics and heredity
Mutualists	two species that benefit one another by enhancing one another's survival, growth, or reproduction
Native species	species that historically lived in a geographic area
Natural resources	physical and biological goods provided by ecosystems
Natural selection	process by which genes of individuals with traits that increase survival and reproduction are more likely to be passed on to subsequent generations
Niche	conditions that a species needs to survive, grow, and reproduce
Non-native species	species that did not occur in a given geographic area historically
Non-renewable resources	natural resources that cannot make more of themselves (e.g. oil, metals)
Novel ecosystems	natural ecosystems that are in a new, self-reinforcing state where restoration to historic ecosystems is not feasible
Nutrients	simple inorganic molecules that contain critical elements that primary producers (typically plants) need for growth (e.g., ammonium which contains nitrogen or phosphate which contains phosphorous)
Ocean acidification	decreasing pH of the world's oceans due to increased carbonic acid formation under higher atmospheric carbon dioxide levels

One health	interdisciplinary approach that links public health and environmental health
Paradigm	worldview including assumptions societies make about the nature of reality
Photosynthesis	cellular and chemical processes that allow certain organisms to produce their own sugars from sunlight, nutrients, carbon dioxide, and water
Planetary boundaries	refers to boundary within which planetary processes occurred during the Holocene (e.g. level of carbon in atmosphere and associated climate regulation)
Population	individuals of the same species that live and interact with one another
Precautionary principle	idea that changes to the environment should be made cautiously because of the unknown nature of effects
Prediction	**If, then** statement that follows logically from a testable hypothesis. **If,** the world works the way I think it does, **then** if I do this, that should happen.
Primary producers	organisms that are able to synthesize complex organic compounds via either photosynthesis or chemosynthesis
Provisioning services	a subgroup of ecosystem services that includes all the ways nature provides us with goods such as food or clean air
Psychological safety	when people feel comfortable to admit failures and take risks
Reactive nitrogen	forms of nitrogen that can easily react chemically with other molecules
Reciprocity	a reciprocal relationship, an exchange of goods and benefits between two parties
Recombination	creation of new varieties of nucleotide sequences (DNA code) due to exchange among chromosomes during meiosis
Reconciliation ecology	approach that seeks to bring nature stewardship and biodiversity conservation to human dominated landscapes
Recycling	reusing materials to create new goods and materials
Regime shift	see ecological regime shift
Regulating service	a subgroup of ecosystem services that includes all the ways nature keeps the environment within optimal conditions for human well-being
Renewable resources	natural resources that are either able to make more of themselves (e.g. trees) or are so abundant that they cannot be depleted (e.g. wind, sunlight)

Reproducibility	the ability to recreate scientific results by redoing a study
Scientific consensus	agreement across the scientific community
Scientific laws	descriptions of consistent patterns and relationships that have been demonstrated by empirical evidence
Scientific method	developing predictions to test hypotheses, evaluating and refining hypotheses in response to results
Scientific theory	a well-tested scientific hypothesis that is supported by a preponderance of evidence (e.g. heliocentric theory)
Specialists	organisms with narrow and highly specific requirements for survival
Speciation	creation of new species as populations become genetically separated from one another and diverge over time
Species	a distinct evolutionary group with genetics and often other characteristics that distinguish it from other groups
Species diversity	diversity of life at the species scale including both the number of species and their relative abundance to one another in ecosystems
Succession	long-term changes in a biological community at a location following an ecological disturbance
Supporting services	a subgroup of ecosystem services that includes all of the basic ecosystem processes that make life on the planet possible
Sustainability	having industries, economies, cultures, and lifestyles such that we live within the earth's limits and leave the planet healthy for future generations
Testable hypothesis	a hypothesis that you can test by gathering empirical evidence
Trophic level	place in the food chain starting with the primary producers (first trophic level) and counting up subsequent steps to higher levels ending with an apex predator
Troposphere	layer of earth's atmosphere closest to earth extending several kilometers above the earth's surface
Wicked problems	systemic level problems that are interconnected, complex, and defy easy solution

Index

Note: Locators in *italic* and **bold** refer to figures and tables respectively, and locators followed by "n" refer to chapter end notes.

For Product Safety Concerns and Information please contact our EU
representative GPSR@taylorandfrancis.com
Taylor & Francis Verlag GmbH, Kaufingerstraße 24, 80331 München, Germany

www.ingramcontent.com/pod-product-compliance
Ingram Content Group UK Ltd.
Pitfield, Milton Keynes, MK11 3LW, UK
UKHW020002200926
475702UK00015B/2433
* 9 7 8 1 0 4 1 2 1 0 7 2 6 *